GCSE
Additional
Science
Higher

Complete Revision and Practice

For the Year 11 exams

Contents

The Periodic Table

Periods

Group 0

	Group 1	Group 2											Group 3	Group 4	Group 5	Group 6	Group 7	
1						1 H Hydrogen 1												4 He Helium 2
2	7 Li Lithium 3	9 Be Beryllium 4											11 B Boron 5	12 C Carbon 6	14 N Nitrogen 7	16 O Oxygen 8	19 F Fluorine 9	20 Ne Neon 10
3	23 Na Sodium 11	24 Mg Magnesium 12											27 Al Aluminium 13	28 Si Silicon 14	31 P Phosphorus 15	32 S Sulfur 16	35.5 Cl Chlorine 17	40 Ar Argon 18
4	39 K Potassium 19	40 Ca Calcium 20	45 Sc Scandium 21	48 Ti Titanium 22	51 V Vanadium 23	52 Cr Chromium 24	55 Mn Manganese 25	56 Fe Iron 26	59 Co Cobalt 27	59 Ni Nickel 28	63.5 Cu Copper 29	65 Zn Zinc 30	70 Ga Gallium 31	73 Ge Germanium 32	75 As Arsenic 33	79 Se Selenium 34	80 Br Bromine 35	84 Kr Krypton 36
5	85 Rb Rubidium 37	88 Sr Strontium 38	89 Y Yttrium 39	91 Zr Zirconium 40	93 Nb Niobium 41	96 Mo Molybdenum 42	99 Tc Technetium 43	101 Ru Ruthenium 44	103 Rh Rhodium 45	106 Pd Palladium 46	108 Ag Silver 47	112 Cd Cadmium 48	115 In Indium 49	119 Sn Tin 50	122 Sb Antimony 51	128 Te Tellurium 52	127 I Iodine 53	131 Xe Xenon 54
6	133 Cs Caesium 55	137 Ba Barium 56	57-71 Lanthanides	178 Hf Hafnium 72	181 Ta Tantalum 73	184 W Tungsten 74	186 Re Rhenium 75	190 Os Osmium 76	192 Ir Iridium 77	195 Pt Platinum 78	197 Au Gold 79	201 Hg Mercury 80	204 Tl Thallium 81	207 Pb Lead 82	209 Bi Bismuth 83	210 Po Polonium 84	210 At Astatine 85	222 Rn Radon 86
7	223 Fr Francium 87	226 Ra Radium 88	89-103 Actinides															

Atomic number →

Published by CGP

From original material by Richard Parsons.

Editors:
Katherine Craig, Helen Ronan, Camilla Simson.

Contributors:
Michael Aicken, Mike Dagless, James Foster, Barbara Mascetti, John Myers, Mike Thompson.

ISBN: 978 1 84146 655 2

With thanks to Mary Falkner for the proofreading.
With thanks to Anna Lupton for the copyright research.

Data used to draw graph on page 52, source developed by the National Center for Health Statistics in collaboration with the National Center for Chronic Disease Prevention and Health Promotion (2000). http://www.cdc.gov/growthcharts

Data used to construct stopping distance diagram on page 192 from the Highway Code.
© Crown Copyright re-produced under the terms of the Click-Use licence.

Every effort has been made to locate copyright holders and obtain permission to reproduce sources. For those sources where it has been difficult to trace the originator of the work, we would be grateful for information. If any copyright holder would like us to make an amendment to the acknowledgements, please notify us and we will gladly update the book at the next reprint. Thank you.

Groovy website: www.cgpbooks.co.uk

Printed by Elanders Ltd, Newcastle upon Tyne.
Jolly bits of clipart from CorelDRAW®

The Scientific Process

Before you get started with the really fun stuff, it's a good idea to understand exactly <u>how</u> the world of science <u>works</u>. Investigate these next few pages and you'll be laughing all day long on results day.

Scientists Come Up with **Hypotheses** — Then **Test** Them

1) Scientists try to <u>explain</u> things. Everything.

2) They start by <u>observing</u> or <u>thinking about</u> something they don't understand — it could be anything, e.g. planets in the sky, a person suffering from an illness, what matter is made of... anything.

3) Then, using what they already know (plus a bit of insight), they come up with a <u>hypothesis</u> — a possible <u>explanation</u> for what they've observed.

4) The next step is to <u>test</u> whether the hypothesis might be <u>right or not</u> — this involves <u>gathering evidence</u> (i.e. <u>data</u> from <u>investigations</u>).

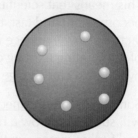

About 100 years ago, scientists hypothesised that atoms looked like this.

5) To gather evidence the scientist uses the hypothesis to make a <u>prediction</u> — a statement based on the hypothesis that can be <u>tested</u> by carrying out <u>experiments</u>.

6) If the results from the experiments match the prediction, then the scientist can be <u>more confident</u> that the hypothesis is <u>correct</u>. This <u>doesn't</u> mean the hypothesis is <u>true</u> though — other predictions based on the hypothesis might turn out to be <u>wrong</u>.

Scientists **Work Together** to Test Hypotheses

After more evidence was gathered, scientists changed their hypothesis to this.

1) Different scientists can look at the <u>same evidence</u> and interpret it in <u>different ways</u>. That's why scientists usually work in <u>teams</u> — they can share their <u>different ideas</u> on how to interpret the data they find.

2) Once a team has come up with (and tested) a hypothesis they all agree with, they'll present their work to the scientific community through <u>journals</u> and <u>scientific conferences</u> so it can be judged — this is called the <u>peer review</u> process.

3) Other scientists then <u>check</u> the team's results (by trying to <u>replicate</u> them) and carry out their own experiments to <u>collect more evidence</u>.

4) If all the experiments in the world back up the hypothesis, scientists start to have a lot of <u>confidence</u> in it. (A hypothesis that is <u>accepted</u> by pretty much every scientist is referred to as a <u>theory</u>.)

5) However, if another scientist does an experiment and the results <u>don't</u> fit with the hypothesis (and other scientists can <u>replicate</u> these results), then the hypothesis is in trouble. When this happens, scientists have to come up with a new hypothesis (maybe a <u>modification</u> of the old explanation, or maybe a completely <u>new</u> one).

A hypothesis is a possible explanation for an observation

If scientists think something is true, they need to produce evidence to convince others — it's all part of <u>testing a hypothesis</u>. One hypothesis might survive these tests, while others won't — it's how things progress. And along the way some hypotheses will be disproved — i.e. shown not to be true.

The Scientific Process

*Scientific Ideas **Change** as **New Evidence** is Found*

1) Scientific explanations are <u>provisional</u> because they only explain the evidence that's <u>currently available</u> — new evidence may come up that can't be explained.

2) This means that scientific explanations <u>never</u> become hard and fast, totally indisputable <u>fact</u>. As <u>new evidence</u> is found (or new ways of <u>interpreting</u> existing evidence are found), hypotheses can <u>change</u> or be <u>replaced</u>.

3) Sometimes, an <u>unexpected observation</u> or <u>result</u> will suddenly throw a hypothesis into doubt and further experiments will need to be carried out. This can lead to new developments that <u>increase</u> our <u>understanding</u> of science.

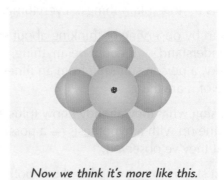

Now we think it's more like this.

*Scientific Developments are **Great**, but they can **Raise Issues***

Scientific <u>knowledge is increased</u> by doing experiments.
And this knowledge leads to <u>scientific developments</u>, e.g. new technologies or new advice.
These developments can create <u>issues</u> though. For example:

Economic issues:

Society <u>can't</u> always <u>afford</u> to do things scientists recommend (e.g. investing heavily in alternative energy sources) without <u>cutting back elsewhere</u>.

Social issues:

Decisions based on scientific evidence affect <u>people</u> — e.g. should fossil fuels be taxed more highly (to invest in alternative energy)? Should alcohol be banned (to prevent health problems)? <u>Would the effect on people's lifestyles be acceptable...</u>

Environmental issues:

<u>Genetically modified crops</u> may help us <u>produce more food</u> — but some people think they could cause <u>environmental problems</u>.

Ethical issues:

There are a lot of things that scientific developments have made possible, but <u>should we do them</u>? E.g. clone humans, develop better nuclear weapons.

It's not all test tubes and explosions

Life can be hard as a scientist. You think you've got it all figured out and then someone comes along with some <u>issues</u>. But it's for the best really — the world would be pretty messed up if no-one ever thought about the issues created by scientific developments (just watch Jurassic Park...).

Quality of Data

The scientific community won't just accept any old results... They've got to be <u>reliable</u> and <u>valid</u>.

Evidence Needs to be **Reliable** (*Repeatable* and *Reproducible*)

<u>RELIABLE</u> means that the data can be <u>repeated</u>, and <u>reproduced</u> by others.

Evidence is only <u>reliable</u> if it can be <u>repeated</u> (during an experiment) AND <u>other scientists can reproduce it too</u> (in other experiments). If it's not reliable, you can't believe it.

<u>EXAMPLE:</u>

In 1998, a scientist claimed that he'd found a link between the MMR vaccine (for measles, mumps and rubella) and autism.

As a result, many parents stopped their children from having the vaccine — which led to a big rise in the number of children catching measles.

However, no other scientist has been able to repeat the results since — they just weren't reliable. Health authorities have now concluded that the vaccine is safe to use.

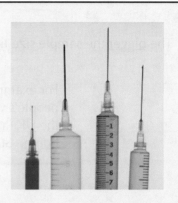

Evidence Also Needs to Be **Valid**

<u>VALID</u> means that the data is <u>reliable</u> AND <u>answers the original question</u>.

<u>EXAMPLE: Do mobile phones cause brain tumours?</u>

- Some studies have found that people who use mobile phones regularly are more likely to develop brain tumours. What they'd actually found was a correlation (relationship) between the variables "use of mobile phones" and "development of brain tumours" — they found that as one changed, so did the other.

- But this evidence is not enough to say that using a mobile phone causes brain tumours, as other explanations might be possible. For example, age, gender and family history can all increase the risk of developing a brain tumour.

- So these studies don't show a definite link and so don't answer the original question.

RRRR — Remember, Reliable means Repeatable and Reproducible

The scientific community <u>won't accept</u> someone's data if it can't be <u>repeated</u> by anyone else. It may sound like a really fantastic new theory, but if there's no other <u>support</u> for it, it just <u>isn't reliable</u>.

Quality of Data

The way evidence is <u>gathered</u> can have a big effect on how <u>trustworthy</u> it is...

*The **Bigger** the **Sample Size** the **Better***

1) Data based on <u>small samples</u> isn't as good as data based on large samples.

> A sample should be <u>representative</u> of the <u>whole</u> <u>population</u> (i.e. it should share as many of the various characteristics in the population as possible) — a small sample can't do that as well.

2) The <u>bigger</u> the sample size the <u>better</u>, but scientists have to be <u>realistic</u> when choosing how big.

> For example, if you were studying how lifestyle affects people's weight it'd be great to study everyone in the UK (a huge sample), but it'd take <u>ages</u> and <u>cost a</u> <u>bomb</u>. Studying a thousand people is more <u>realistic</u>.

*Don't Always **Believe** What You're Being **Told Straight Away***

1) People who want to make a point might <u>present data</u> in a <u>biased way</u>, e.g. by overemphasising a relationship in the data. (Sometimes <u>without knowing</u> they're doing it.)

2) And there are all sorts of reasons <u>why</u> people might <u>want</u> to do this — for example, <u>companies</u> might want to 'big up' their products. Or make impressive safety claims.

3) If an investigation is done by a team of <u>highly-regarded scientists</u> it's sometimes taken <u>more seriously</u> than evidence from <u>less well known scientists</u>.

4) But having experience, authority or a fancy qualification <u>doesn't</u> necessarily mean the evidence is <u>good</u> — the only way to tell is to look at the evidence scientifically (e.g. is it reliable, valid, etc.).

Things are not always what they seem

No matter <u>what</u> you're reading or <u>who</u> it's written by you've always got to be really careful about what you <u>believe</u>. Ask yourself whether the <u>sample</u> is a decent <u>size</u> and check whether the author has anything to gain from what's written. For example, an article on the magical fat-busting power of spinach written by the country's leading spinach grower may not be all it seems. <u>Don't be fooled</u>.

Limits of Science

Science can give us <u>amazing things</u> — cures for diseases, space travel, heated toilet seats...
But science has its <u>limitations</u> — there are questions that it just can't answer.

Some Questions Are *Unanswered*...

Some questions are <u>unanswered</u> — we <u>don't know everything</u> and we <u>never will</u>.
We'll find out <u>more</u> as new hypotheses are suggested and more experiments are done,
but there'll <u>always</u> be stuff we don't know.

> For example, we don't know what the <u>exact impacts</u>
> of <u>global warming</u> are going to be. At the moment
> scientists don't all agree on the answers because
> there <u>isn't enough</u> reliable and valid <u>evidence</u>.

...*Others are* **Unanswerable**

1) Then there's the other type... questions that all the experiments in the world <u>won't</u> help us answer
    — the "<u>Should we be doing this at all?</u>" type questions. There are always two sides...

2) Take <u>embryo screening</u> (which allows you to choose an embryo with particular characteristics).
    It's <u>possible</u> to do it — but does that mean we <u>should</u>?

3) Different people have <u>different opinions</u>. For example...

- Some people say it's <u>good</u>... couples whose <u>existing</u> child
  needs a <u>bone marrow transplant</u>, but who can't find
  a donor, will be able to have <u>another</u> child selected for its
  <u>matching</u> bone marrow. This would <u>save</u> the life of their
  first child — and if they <u>want</u> another child anyway...
  where's the harm?

- Other people say it's <u>bad</u>... they say it could have
  serious effects on the <u>new child</u>. In the above example,
  the new child might feel <u>unwanted</u> — thinking they
  were only brought into the world to help <u>someone else</u>.
  And would they have the right to <u>refuse</u> to donate their
  bone marrow (as anyone else would)?

4) The question of whether something is <u>morally</u> or <u>ethically</u> right or wrong <u>can't be answered</u> by more
    <u>experiments</u> — there is <u>no "right" or "wrong" answer</u>.

5) The best we can do is get a <u>consensus</u> from society — a <u>judgement</u> that <u>most people</u> are more or less
    happy to live by. <u>Science</u> can provide <u>more information</u> to help people make this judgement, and the
    judgement might <u>change</u> over time. But in the end it's up to <u>people</u> and their <u>conscience</u>.

To answer or not to answer, that is the question

It's official — <u>no-one</u> knows everything. Your teacher/mum/annoying older sister (delete as applicable)
might think and act as if they know it all, but sadly they <u>don't</u>. So in reality you know one thing they
don't — which clearly makes you more intelligent and generally far superior in every way. Possibly.

Planning Investigations

The next few pages show how <u>investigations</u> should be carried out — by both <u>scientists</u> and <u>you</u>.

In a *Fair Test* You Have to *Control the Variables*

1) In a lab experiment you usually <u>change one variable</u> and <u>measure</u> how it affects the <u>other variable</u>.

> <u>EXAMPLE</u>: you might change only the temperature of an enzyme-controlled reaction and measure how it affects the rate of reaction.

2) To make it a fair test <u>everything else</u> that could affect the results should <u>stay the same</u> (otherwise you can't tell if the thing that's being changed is affecting the results or not — the data won't be reliable or valid).

> <u>EXAMPLE continued</u>: you need to keep the pH the same, otherwise you won't know if any change in the rate of reaction is caused by the change in temperature, or the change in pH.

3) The variable that you <u>change</u> is called the <u>independent</u> variable.

4) The variable that's <u>measured</u> is called the <u>dependent</u> variable.

5) The variables that you <u>keep the same</u> are called <u>control</u> variables.

> <u>EXAMPLE continued</u>:
> Independent = temperature
> Dependent = rate of reaction
> Control = pH

6) Because you can't always control all the variables, you often need to use a <u>control experiment</u> — an experiment that's kept under the <u>same conditions</u> as the rest of the investigation, but doesn't have anything done to it. This is so that you can see what happens when you don't change anything at all.

Experiments Must be *Safe*

1) Part of planning an investigation is making sure that it's <u>safe</u>.

2) A <u>hazard</u> is something that can <u>potentially cause harm</u>.

3) There are lots of <u>hazards</u> you could be faced with during an investigation, e.g. <u>radiation</u>, <u>electricity</u>, <u>gas</u>, <u>chemicals</u> and <u>fire</u>.

4) You should always make sure that you <u>identify</u> all the hazards that you might encounter.

5) You should also come up with ways of <u>reducing the risks</u> from the hazards you've identified.

6) One way of doing this is to carry out a <u>risk assessment</u>:

> For an experiment involving a <u>Bunsen burner</u>, the risk assessment might be something like this:

> <u>Hazard</u>: Bunsen burner is a fire risk.
> <u>Precautions</u>:
> • Keep flammable chemicals away from the Bunsen.
> • Never leave the Bunsen unattended when lit.
> • Always turn on the yellow safety flame when not in use.

`DANGER`
HARD HATS
AND
SAFETY BOOTS
MUST BE WORN
ON THIS SITE

Hazard: revision boredom. Precaution: use CGP books

Labs are dangerous places — you need to know the <u>hazards</u> of what you're doing <u>before you start</u>.

Collecting Data

There are a few things that can be done to make sure that you get the best results you possibly can.

The **Equipment** Used has to be **Right for the Job**

1) The measuring equipment you use has to be sensitive enough to accurately measure the chemicals you're using.

Accurate data is data that's close to the true value — see the next page.

E.g. if you need to measure out 11 ml of a liquid, you'll need to use a measuring cylinder that can measure to 1 ml, not 5 or 10 ml.

2) The smallest change a measuring instrument can detect is called its RESOLUTION. E.g. some mass balances have a resolution of 1 g and some have a resolution of 0.1 g.

3) Also, equipment needs to be calibrated so that your data is more accurate. E.g. mass balances need to be set to zero before you start weighing things.

Trial Runs give you the **Range** and **Interval** of **Variable Values**

1) Before you carry out an experiment, it's a good idea to do a trial run first — a quick version of your experiment.

2) Trial runs help you work out whether your plan is right or not — you might decide to make some changes after trying out your method.

3) Trial runs are used to figure out the range of variable values used (the upper and lower limit).

Enzyme-controlled reaction example
from previous page continued:
You might do trial runs at 10, 20, 30, 40 and
50 °C. If there was no reaction at 10 or 50 °C,
you might narrow the range to 20-40 °C.

4) And they're used to figure out the interval (gaps) between the values too.

Enzyme-controlled reaction example continued:
If using 10 °C intervals gives you a big change
in rate of reaction you might decide to use 5 °C
intervals, e.g. 20, 25, 30, 35...

5) Trial runs can also help you figure out how many times the experiment has to be repeated to get reliable results. E.g. if you repeat it two times and the results are all similar, then two repeats is enough.

8

Collecting Data

Data Should be as **Reliable**, **Accurate** and **Precise** as Possible

Reliable

1) When carrying out an investigation, you can <u>improve</u> the reliability of your results (see p.3) by <u>repeating</u> the readings and calculating the mean (average, see page 9). You should repeat readings at least <u>twice</u> (so that you have at least <u>three</u> readings to calculate an average result).

2) To make sure your results are reliable you can cross check them by taking a <u>second set of readings</u> with <u>another instrument</u> (or a <u>different observer</u>).

3) Checking your results match with <u>secondary sources</u>, e.g. studies that other people have done, also increases the reliability of your data.

Accurate

1) You should always make sure that your results are <u>accurate</u>. Really accurate results are those that are <u>really close</u> to the <u>true answer</u>.

2) You can get accurate results by doing things like making sure the <u>equipment</u> you're using is <u>sensitive enough</u> (see previous page), and by recording your data to a suitable <u>level of accuracy</u>. For example, if you're taking digital readings of something, the results will be more accurate if you include at least a couple of decimal places instead of rounding to whole numbers.

Precise

1) You should also always make sure your results are <u>precise</u>.

2) Precise results are ones where the data is <u>all really close</u> to the <u>mean</u> (i.e. not spread out).

You Can Check For **Mistakes Made** When **Collecting Data**

1) When you've collected all the results for an experiment, you should have a look to see if there are any results that <u>don't seem to fit</u> in with the rest.

2) Most results vary a bit, but any that are totally different are called <u>anomalous results</u>.

> Barry's results
> measurement 1 — 2.34 cm   measurement 2 — 5.67 cm
> measurement 3 — 2.35 cm   measurement 4 — 2.33 cm

Measurement 2 is an anomalous result — it's totally different from the rest.

3) They're <u>caused</u> by <u>human errors</u>, e.g. by a mistake when measuring.

4) The only way to stop them happening is by taking all your measurements as <u>carefully</u> as possible.

5) If you ever get any anomalous results, you should investigate them to try to <u>work out what happened</u>. If you can work out what happened (e.g. you measured something wrong) you can <u>ignore</u> them when processing your results.

Results need to be reliable, accurate and precise

All this stuff is really important — without <u>good quality</u> data an investigation will be totally <u>meaningless</u>. Get to know this page and your data will be the envy of the whole scientific community.

Organising and Processing Data

The fun doesn't stop once the data's been collected — it then needs to be organised and processed...

Data Needs to be Organised

1) Data that's been collected needs to be organised so it can be processed later on.
2) Tables are dead useful for organising data.
3) When drawing tables you should always make sure that each column has a heading and that you've included the units.

Test tube	Result (ml)	Repeat 1 (ml)	Repeat 2 (ml)
A	28	37	32
B	47	51	60
C	68	72	70

4) Annoyingly, tables are about as useful as a chocolate teapot for showing patterns or relationships in data. You need to use some kind of graph or mathematical technique for that...

Data Can be Processed Using a Bit of Maths

1) Raw data generally just ain't that useful. You usually have to process it in some way.
2) A couple of the most simple calculations you can perform are the mean and the range:

Mean: the average

To calculate the mean ADD TOGETHER all the data values and DIVIDE by the total number of values. You usually do this to get a single value from several repeats of your experiment.

Test tube	Result (ml)	Repeat 1 (ml)	Repeat 2 (ml)	Mean (ml)	Range (ml)
A	28	37	32	(28 + 37 + 32) ÷ 3 = 32.3	37 – 28 = 9
B	47	51	60	(47 + 51 + 60) ÷ 3 = 52.7	60 – 47 = 13
C	68	72	70	(68 + 72 + 70) ÷ 3 = 70.0	72 – 68 = 4

Range: how spread out the data is

To calculate the range find the LARGEST number and SUBTRACT the SMALLEST number. You usually do this to check the accuracy and reliability of the results — the greater the spread of the data, the lower the accuracy and reliability.

Processing data requires a teeny bit of maths — don't panic

This data stuff is pretty straightforward — but it's really important. Different measurements scattered on random bits of paper at the bottom of your bag just ain't gonna cut it. Get your ruler out, draw a lovely table, pop your data in, calculate the mean and the range and breathe... Much better.

Presenting Data

Data has to be <u>presented</u>, and graphs are just about the best way of doing it...

Different Types of *Data* Should be *Presented* in *Different Ways*

1) Once you've carried out an investigation, you'll need to <u>present</u> your data so that it's easier to see <u>patterns</u> and <u>relationships</u> in the data.

2) Different types of investigations give you <u>different types</u> of data, so you'll always have to <u>choose</u> what the best way to present your data is.

Pie charts can be used to present the same sort of data as bar charts. They're mostly used when the data is in percentages or fractions though.

Bar Charts

If the independent variable is <u>categoric</u> (comes in distinct categories, e.g. blood types, metals) you should use a <u>bar chart</u> to display the data. You also use them if the independent variable is <u>discrete</u> (the data can be counted in chunks, where there's no in-between value, e.g. number of people is discrete because you can't have half a person).

There are some <u>golden rules</u> you need to follow for <u>drawing</u> bar charts:

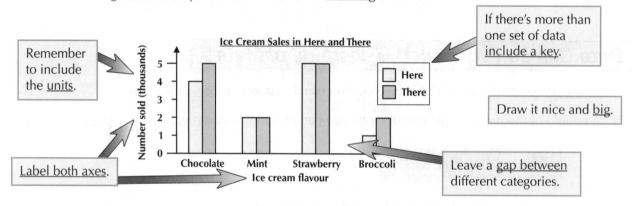

Remember to include the <u>units</u>.

If there's more than one set of data <u>include a key</u>.

Draw it nice and <u>big</u>.

<u>Label both axes</u>.

Leave a <u>gap between</u> different categories.

Ice Cream Sales in Here and There

Number sold (thousands)

Key: Here / There

Chocolate, Mint, Strawberry, Broccoli

Ice cream flavour

Line Graphs

If the independent variable is <u>continuous</u> (numerical data that can have any value within a range, e.g. length, volume, temperature) you should use a <u>line graph</u> to display the data.

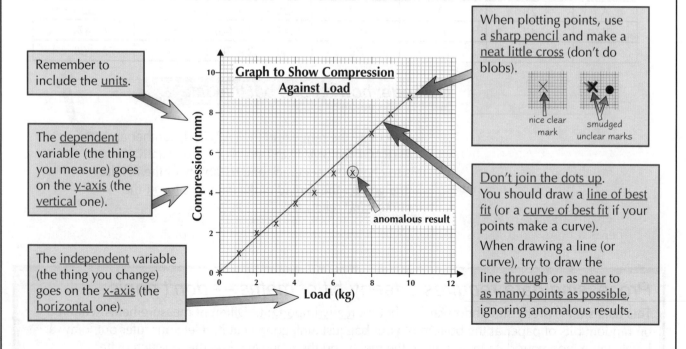

Remember to include the <u>units</u>.

The <u>dependent</u> variable (the thing you measure) goes on the <u>y-axis</u> (the <u>vertical</u> one).

The <u>independent</u> variable (the thing you change) goes on the <u>x-axis</u> (the <u>horizontal</u> one).

Graph to Show Compression Against Load

Compression (mm)

Load (kg)

anomalous result

When plotting points, use a <u>sharp pencil</u> and make a <u>neat little cross</u> (don't do blobs).

nice clear mark

smudged unclear marks

<u>Don't join the dots up</u>. You should draw a <u>line of best fit</u> (or a <u>curve of best fit</u> if your points make a curve).

When drawing a line (or curve), try to draw the line <u>through</u> or as <u>near</u> to <u>as many points as possible</u>, ignoring anomalous results.

Interpreting Data

Once you've drawn your graph (using all the tips on the previous page) you need to be able to <u>understand</u> what it's <u>telling you</u>. That's where this page comes in handy.

Line Graphs Can Show *Relationships* in *Data*

1) Line graphs are great for showing relationships <u>between two variables</u> (just like other graphs).

2) Here are some of the different types of <u>correlation</u> (relationship) shown on line graphs:

Positive Correlation

As one variable <u>increases</u> the other <u>increases</u>.

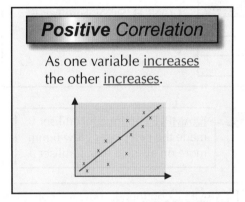

Inverse (Negative) Correlation

As one variable <u>increases</u> the other <u>decreases</u>.

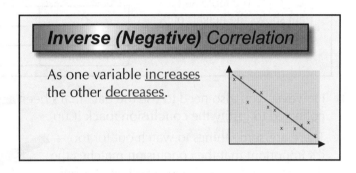

No Correlation

There's <u>no relationship</u> between the two variables.

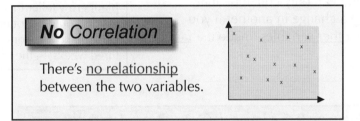

Linear

The graph is a <u>straight line</u>.

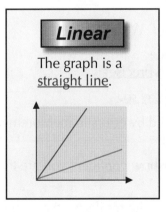

Directly Proportional

The graph is a <u>straight line</u> where both variables increase (or decrease) in the <u>same ratio</u>.

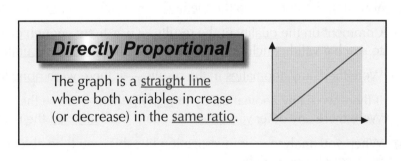

3) You've got to be careful not to <u>confuse correlation</u> with <u>cause</u> though. A <u>correlation</u> just means that there's a <u>relationship</u> between two variables. It <u>doesn't always mean</u> that the change in one variable is <u>causing</u> the change in the other.

4) There are <u>three possible reasons</u> for a correlation. It could be down to <u>chance</u>, it could be that there's a <u>third variable</u> linking the two things, or it might actually be that one variable is <u>causing</u> the other to change.

A correlation is a relationship between sets of data

Wow. What a snazzy page. But don't get distracted by all those crazy lines and crosses — <u>interpreting</u> data's dead important so make sure you know how to do it properly.

Concluding and Evaluating

At the end of an investigation, the <u>conclusion</u> and <u>evaluation</u> are waiting. Don't worry, they won't bite.

A **Conclusion** is a **Summary** of What You've **Learnt**

1) Once all the data's been collected, presented and analysed, an investigation will always involve coming to a <u>conclusion</u>.

2) Drawing a conclusion can be quite straightforward — just <u>look at your data</u> and <u>say what pattern you see</u>.

<u>EXAMPLE</u>: The table on the right shows the heights of pea plant seedlings grown for three weeks with different fertilisers.

Fertiliser	Mean growth (mm)
A	13.5
B	19.5
No fertiliser	5.5

<u>CONCLUSION</u>: Fertiliser <u>B</u> makes <u>pea plant</u> seedlings grow taller over a <u>three week</u> period than fertiliser A.

3) However, you also need to use the data that's been <u>collected</u> to <u>justify</u> the conclusion (back it up).

<u>EXAMPLE continued</u>: Fertiliser B made the pea plants grow 6 mm more on average than fertiliser A.

4) There are some things to watch out for too — it's important that the conclusion <u>matches the data</u> it's based on and <u>doesn't go any further</u>.

5) Remember not to <u>confuse correlation</u> and <u>cause</u> (see previous page). You can only conclude that one variable is <u>causing</u> a change in another if you have controlled all the <u>other variables</u> (made it a <u>fair test</u>).

<u>EXAMPLE continued</u>: You can't conclude that fertiliser B makes <u>any other type of plant</u> grow taller than fertiliser A — the results could be totally different. Also, you can't make any conclusions <u>beyond</u> the three weeks — the plants could <u>drop dead</u>.

Evaluations — Describe **How** it Could be **Improved**

An evaluation is a <u>critical analysis</u> of the whole investigation.

1) You should comment on the <u>method</u> — was the <u>equipment suitable</u>? Was it a <u>fair test</u>?

2) Comment on the <u>quality</u> of the <u>results</u> — was there <u>enough evidence</u> to reach a valid <u>conclusion</u>? Were the results <u>reliable</u>, <u>accurate</u> and <u>precise</u>?

3) Were there any <u>anomalies</u> in the results — if there were <u>none</u> then <u>say so</u>.

4) If there were any anomalies, try to <u>explain</u> them — were they caused by <u>errors</u> in measurement? Were there any other <u>variables</u> that could have <u>affected</u> the results?

5) When you analyse your investigation like this, you'll be able to say how <u>confident</u> you are that your conclusion is <u>right</u>.

6) Then you can suggest any <u>changes</u> that would <u>improve</u> the quality of the results, so that you could have <u>more confidence</u> in your conclusion. For example, you might suggest changing the way you controlled a variable, or changing the interval of values you measured.

7) You could also make more <u>predictions</u> based on your conclusion, then <u>further experiments</u> could be carried out to test them.

8) When suggesting improvements to the investigation, always make sure that you say <u>why</u> you think this would make the results <u>better</u>.

An experiment must have a conclusion and an evaluation

I know it doesn't seem very nice, but writing about where you went <u>wrong</u> is an important skill — it shows you've got a really good understanding of what the investigation was <u>about</u>.

Cells

All living things are made of <u>cells</u>. When someone first peered down a microscope at a slice of cork and drew the <u>boxes</u> they saw, little did they know that they'd seen the <u>building blocks</u> of <u>every organism on the planet</u>.

Animal and Plant Cells Have Certain Features in Common

Most <u>animal</u> and <u>plant</u> cells have the following parts:

Animal cell

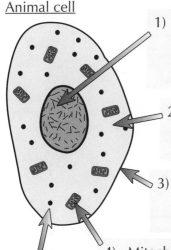

1) <u>Nucleus</u> — contains <u>DNA</u> (see page 40). DNA contains the instructions for making <u>proteins</u>, e.g. the <u>enzymes</u> used in the chemical reactions of <u>respiration</u> (in animal and plant cells, see pages 28-30) and <u>photosynthesis</u> (in plant cells only, see page 76).

2) <u>Cytoplasm</u> — gel-like substance where proteins like <u>enzymes</u> (see page 20) are made. Some <u>enzyme-controlled reactions</u> take place in the cytoplasm, e.g. the reactions of <u>anaerobic respiration</u> (see page 30).

3) <u>Cell membrane</u> — holds the cell together and controls what goes <u>in</u> and <u>out</u>. It lets <u>gases</u> and <u>water</u> pass through freely while acting as a <u>barrier</u> to other <u>chemicals</u>.

4) <u>Mitochondria</u> — these are where the <u>enzymes</u> needed for the reactions of <u>aerobic respiration</u> (see page 28) are found, and where the reactions take place.

5) <u>Ribosomes</u> — these are where <u>proteins</u> are made in the cell.

Plant Cells have Some Extra Features

Plant cells also have a few <u>extra</u> things that animal cells <u>don't</u> have:

1) Rigid <u>cell wall</u> — made of <u>cellulose</u>. It <u>supports</u> the cell and strengthens it.

Plant cell

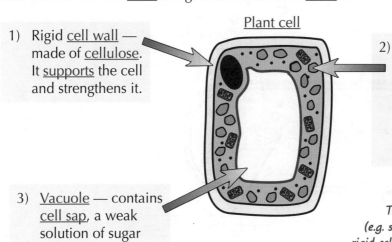

2) <u>Chloroplasts</u> — these are where the reactions for <u>photosynthesis</u> take place. They contain a green substance called <u>chlorophyll</u> and the <u>enzymes</u> needed for photosynthesis.

3) <u>Vacuole</u> — contains <u>cell sap</u>, a weak solution of sugar and salts.

The cells of algae (e.g. seaweed) also have a rigid cell wall and chloroplasts.

There's quite a bit to learn in biology — but that's life, I guess...

On this page are a <u>typical animal cell</u> and a <u>typical plant cell</u>. They have a lot of the <u>same bits</u>, but plant cells have some <u>extra bits</u> too. Strangely enough, not all plant or animal cells are the same — they have different <u>structures</u> and <u>produce</u> different substances depending on the <u>job</u> they do.

Cells

Two types of cell coming up — <u>yeast cells</u> and <u>bacterial cells</u>...

Yeast *is a* **Single-Celled** *Organism*

1) Yeast is a <u>microorganism</u>.
2) A yeast cell has a <u>nucleus</u>, <u>cytoplasm</u>, and a <u>cell membrane</u> surrounded by a <u>cell wall</u>.

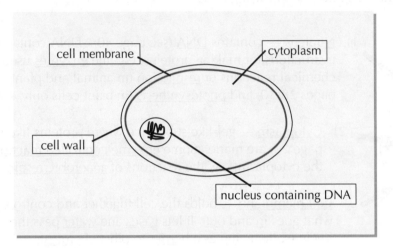

Bacteria Have a **Simple Cell Structure**

1) <u>Bacterial</u> cells are a bit different to plant, animal and yeast cells.
2) They <u>don't</u> have a <u>nucleus</u>. They have a <u>circular molecule of DNA</u> which floats around in the cytoplasm.
3) They <u>don't</u> have <u>mitochondria</u> either, but they <u>can</u> still <u>respire aerobically</u>.

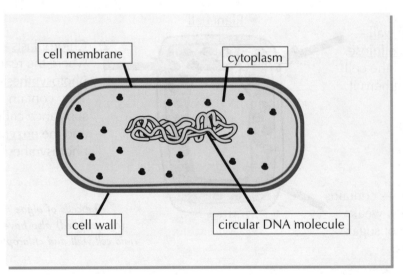

Both yeast and bacteria are single-celled microorganisms

On this page is a typical <u>yeast cell</u> and a typical <u>bacterial cell</u>. And if you haven't already seen enough cells, there are even more coming up on the next couple of pages.

Specialised Cells

Pages 13 and 14 show the structure of some typical cells. However, most cells are <u>specialised</u> for their specific function, so their structure can vary...

1) *Palisade Leaf Cells* Are Adapted for *Photosynthesis*

1) Packed with <u>chloroplasts</u> for <u>photosynthesis</u>. More of them are crammed at the <u>top</u> of the cell — so they're nearer the <u>light</u>.

2) <u>Tall</u> shape means a lot of <u>surface area</u> exposed down the side for <u>absorbing CO$_2$</u> from the air in the leaf.

3) <u>Thin</u> shape means that you can pack loads of them in at the top of a leaf.

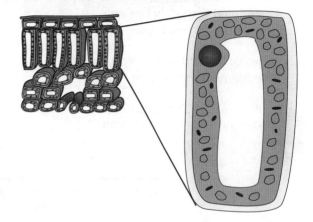

Palisade leaf cells are grouped together at the top of the leaf where most of the <u>photosynthesis</u> (see page 76) happens.

2) *Guard Cells* Are Adapted to *Open and Close Pores*

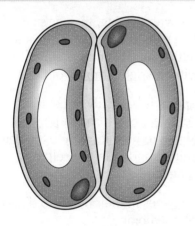

1) Special kidney shape which <u>opens</u> and <u>closes</u> the <u>stomata</u> (pores) in a leaf.

2) When the plant has <u>lots</u> of water the guard cells fill with it and go plump and <u>turgid</u>. This makes the stomata <u>open</u> so <u>gases</u> can be exchanged for <u>photosynthesis</u>.

3) When the plant is <u>short</u> of water, the guard cells lose water and become <u>flaccid</u>, making the stomata <u>close</u>. This helps stop too much water vapour <u>escaping</u>.

4) <u>Thin</u> outer walls and <u>thickened</u> inner walls make the opening and closing work.

5) They're also <u>sensitive to light</u> and <u>close at night</u> to save water without losing out on photosynthesis.

Guard cells are therefore adapted to their function of allowing <u>gas exchange</u> and <u>controlling water loss</u> within a <u>leaf</u>.

Specialised Cells

3) Red Blood Cells Are Adapted to Carry Oxygen

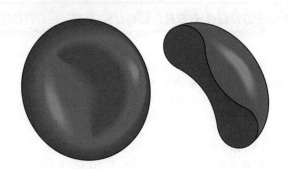

1) <u>Concave</u> shape gives a big <u>surface area</u> for absorbing <u>oxygen</u>. It also helps them pass <u>smoothly</u> through <u>capillaries</u> to reach body cells.

2) They're packed with <u>haemoglobin</u> — the pigment that absorbs the oxygen.

3) They have <u>no nucleus</u>, to leave even more room for haemoglobin.

Red blood cells are an important part of the <u>blood</u>.

4) Sperm and Egg Cells Are Specialised for Reproduction

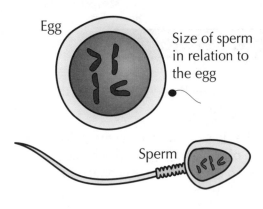

Egg

Size of sperm in relation to the egg

Sperm

1) The main functions of an <u>egg cell</u> are to carry the female DNA and to <u>nourish</u> the developing embryo in the early stages. The egg cell contains huge <u>food reserves</u> to feed the embryo.

2) When a <u>sperm</u> fuses with the egg, the egg's <u>membrane</u> instantly <u>changes</u> its structure to stop any more sperm getting in. This makes sure the offspring end up with the <u>right amount</u> of DNA.

3) The function of a <u>sperm</u> is basically to get the <u>male DNA</u> to the <u>female DNA</u>. It has a <u>long tail</u> and a <u>streamlined head</u> to help it <u>swim</u> to the egg. There are a lot of <u>mitochondria</u> in the cell to provide the <u>energy</u> needed.

4) Sperm also carry <u>enzymes</u> in their heads to digest through the egg cell membrane.

Sperm and eggs are very important cells in <u>reproduction</u>.

Cells have the same basic bits but are specialised for their function

These cells all have all the bits shown on page 13, even though they look completely different and do <u>totally different jobs</u>. Apart from red blood cells — which, for example, don't have a nucleus.

Cell Organisation

How, you might wonder, does having all these specialised cells mean you end up with a working human... the answer's organisation.

Large Multicellular Organisms are Made Up of Organ Systems

1) As you know from the previous pages, specialised cells carry out a particular function.
2) The process by which cells become specialised for a particular job is called differentiation.

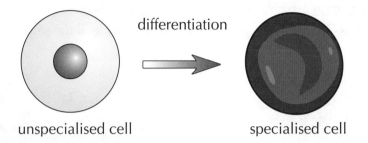

differentiation

unspecialised cell   specialised cell

3) Differentiation occurs during the development of a multicellular organism.
4) These specialised cells form tissues, which form organs, which form organ systems (see page 18).
5) Large multicellular organisms (e.g. humans) have different systems inside them for exchanging and transporting materials.

Similar Cells are Organised into Tissues

1) A tissue is a group of similar cells that work together to carry out a particular function.
2) It can include more than one type of cell.
3) In mammals (like humans), examples of tissues include:

- Muscular tissue, which contracts (shortens) to move whatever it's attached to.
- Glandular tissue, which makes and secretes chemicals like enzymes and hormones.
- Epithelial tissue, which covers some parts of the body, e.g. the inside of the gut.

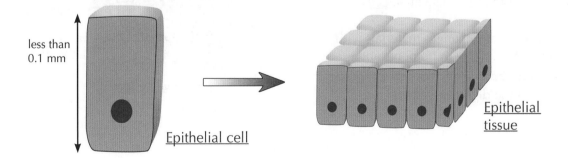

less than 0.1 mm

Epithelial cell

Epithelial tissue

Cell Organisation

We left off at <u>tissues</u> on the previous page — next up is how they're organised...

Tissues are Organised into Organs

1) An <u>organ</u> is a group of <u>different tissues</u> that work together to perform a certain <u>function</u>.

2) For example, the <u>stomach</u> is an organ made of these tissues:

- <u>Muscular tissue</u>, which moves the stomach wall to <u>churn up the food</u>.
- <u>Glandular tissue</u>, which makes <u>digestive juices</u> to digest food.
- <u>Epithelial tissue</u>, which covers the <u>outside</u> and <u>inside</u> of the stomach.

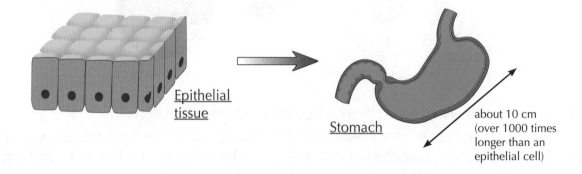

<u>Epithelial tissue</u>

<u>Stomach</u>

about 10 cm (over 1000 times longer than an epithelial cell)

Organs are Organised into Organ Systems

An <u>organ system</u> is a <u>group of organs</u> working together to perform a particular <u>function</u>. For example, the <u>digestive system</u> (found in humans and mammals) <u>breaks down food</u> and is made up of these organs:

1) <u>Glands</u> (e.g. the <u>pancreas</u> and <u>salivary glands</u>), which produce <u>digestive juices</u>.

2) The <u>stomach</u> and <u>small intestine</u>, which <u>digest</u> food.

3) The <u>liver</u>, which produces <u>bile</u>.

4) The <u>small intestine</u>, which <u>absorbs</u> soluble <u>food</u> molecules.

5) The <u>large intestine</u>, which <u>absorbs water</u> from undigested food, leaving <u>faeces</u>.

The digestive system <u>exchanges materials</u> with the <u>environment</u> by <u>taking in nutrients</u> and <u>releasing substances</u> such as bile.

There's more on the digestive system on pages 23-24.

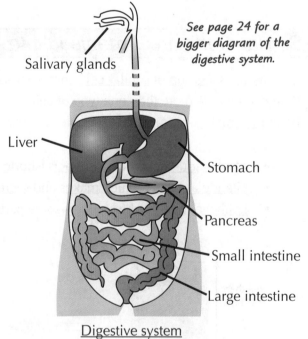

See page 24 for a bigger diagram of the digestive system.

Salivary glands

Liver

Stomach

Pancreas

Small intestine

Large intestine

<u>Digestive system</u>

Remember — cells, tissues, organs, organ systems...

OK, so from the last couple of pages you know that <u>cells</u> are organised into <u>tissues</u>, the tissues into <u>organs</u>, the organs into <u>organ systems</u> and the organ systems into a whole <u>organism</u>.

Warm-Up and Exam Questions

So, hopefully you've read the last six pages. But could you cope if a question on cells came up in an exam? With amazing new technology we can simulate that very situation....

Warm-Up Questions

1) Give three ways in which animal cells are different from plant cells.
2) What does a yeast cell contain that a bacteria cell does not contain?
3) Give two features of bacterial cell.
4) (a) What is the function of a red blood cell?
   (b) Describe two ways in which a red blood cell is adapted to its function.
5) (a) What is the main function of a sperm cell?
   (b) Give one feature of a sperm cell. Explain how this feature helps it to serve its function.
6) What is an organ system?

Exam Questions

1   The diagram shows a palisade cell from a leaf.

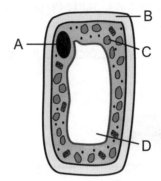

(a) Which label points to a chloroplast?

(1 mark)

(b) Name the green substance present in chloroplasts.

(1 mark)

(c) Apart from having chloroplasts, suggest one other way in which a palisade cell is adapted for photosynthesis.

(1 mark)

2   (a) Complete the following paragraph:

Guard cells are adapted to open and close pores. They can be found in ....

and help to control water loss within plants. When there is lots of water, the guard

cells fill up. This causes them to go .... The .... then

open so that .... can be exchanged for ....

(5 marks)

(b) What happens to guard cells when the plant is short of water?

(1 mark)

(c) Explain why the plant's guard cells go flaccid at night.

(2 marks)

Enzymes

Chemical reactions are what make you work. And enzymes are what make them work.

Enzymes Are Catalysts Produced by Living Things

1) Living things have thousands of different chemical reactions going on inside them all the time.

2) These reactions need to be carefully controlled — to get the right amounts of substances.

3) You can usually make a reaction happen more quickly by raising the temperature.
This would speed up the useful reactions but also the unwanted ones too... not good.
There's also a limit to how far you can raise the temperature inside a living creature before
its cells start getting damaged.

4) So... living things produce enzymes that act as biological catalysts. Enzymes reduce the need
for high temperatures and we only have enzymes to speed up the useful chemical reactions
in the body.

> A CATALYST is a substance which INCREASES the speed of a reaction,
> without being CHANGED or USED UP in the reaction.

5) Enzymes are all proteins and all proteins are made up of chains of amino acids.
These chains are folded into unique shapes, which enzymes need to do their jobs (see below).

6) As well as catalysts, proteins act as structural components of tissues (e.g. muscles), hormones
and antibodies.

Enzymes are Very Specific

1) Chemical reactions usually involve things either being split apart or joined together.

2) A substrate is a molecule that is changed in a reaction.

3) Every enzyme molecule has an active site — the part where a substrate joins on to the enzyme.

4) Enzymes are really picky — they usually only speed up one reaction. This is because, for an
enzyme to work, a substrate has to be the correct shape to fit into the active site.

5) This is called the 'lock and key' model, because the substrate fits into the enzyme just like a key
fits into a lock.

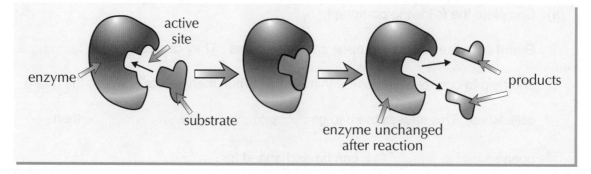

Enzymes speed up chemical reactions

Just like you've got to have the correct key for a lock, you've got to have the right substrate
for an enzyme. If the substrate doesn't fit, the enzyme won't catalyse the reaction...

Enzymes

Enzymes are clearly very clever, but they're <u>not</u> very versatile. They need just the right <u>conditions</u> if they're going to work properly.

Enzymes Need the Right Temperature...

1) Changing the <u>temperature</u> changes the <u>rate</u> of an enzyme-catalysed reaction.

2) Like with any reaction, a higher temperature <u>increases</u> the rate at first.

3) But if it gets <u>too hot</u>, some of the <u>bonds</u> holding the enzyme together <u>break</u>. This destroys the enzyme's <u>special shape</u> and so it won't work any more. It's said to be <u>denatured</u>.

4) Enzymes in the <u>human body</u> normally work best at around <u>37 °C</u>.

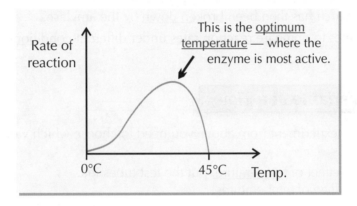

...and the Right pH

1) The <u>pH</u> also affects enzymes. If it's too high or too low, the pH interferes with the <u>bonds</u> holding the enzyme together.

2) This changes the shape and <u>denatures</u> the enzyme.

3) All enzymes have an <u>optimum pH</u> that they work best at. It's often <u>neutral pH 7</u>, but <u>not always</u> — e.g. <u>pepsin</u> is an enzyme used to break down <u>proteins</u> in the <u>stomach</u>. It works best at <u>pH 2</u>, which means it's well-suited to the <u>acidic conditions</u> there.

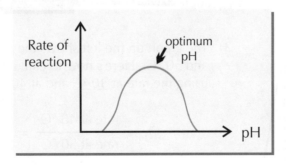

Enzymes work best at the right temperature and pH

The <u>optimum temperature</u> for most human enzymes is around <u>normal body temperature</u>.
And <u>stomach enzymes</u> work best at <u>low pH</u>, but the enzymes in your <u>small intestine</u> like <u>high pH</u>.

Enzyme-Controlled Reactions

You can <u>measure</u> how a variable, e.g. temperature, affects the rate of an <u>enzyme-controlled reaction</u>. Goggles at the ready.

Measuring the *Rate* of an *Enzyme-Controlled* Reaction — *Method*

1) You can measure the rate of a reaction by using <u>amylase</u> as the <u>enzyme</u> and <u>starch</u> as the <u>substrate</u>.

2) <u>Amylase</u> catalyses the breakdown of <u>starch</u>, so you can <u>time</u> how long it takes for the <u>starch</u> to <u>disappear</u>.

3) To do this, regularly take a <u>drop</u> of the amylase and starch mixture, and put it onto a drop of <u>iodine solution</u> on a spotting tile. Record the colour change — it'll turn <u>blue-black</u> if <u>starch</u> is present. Note the <u>time</u> when the iodine solution <u>no longer</u> turns blue-black — the starch has then been <u>broken down</u> by the amylase.

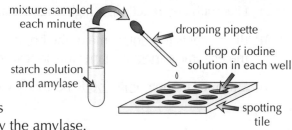

mixture sampled each minute

dropping pipette

drop of iodine solution in each well

starch solution and amylase

spotting tile

4) You can use the times to <u>compare reaction rates</u> under different <u>conditions</u> — see below.

Change One *Variable* at a Time

In the amylase/starch experiment from above you need to choose which variable to change. For example:

1) to investigate the effect of <u>temperature</u>, put the test tubes into <u>water baths</u> at a range of temperatures.

2) to investigate the effect of <u>pH</u>, use a range of different <u>pH buffers</u>.

3) to investigate the effect of <u>substrate concentration</u>, vary the initial <u>concentrations</u> of the <u>starch solutions</u>.

Remember to keep all the variables you're not investigating constant, e.g. use the same amylase concentration each time.

Q_{10} *Values* Show How *Reaction Rate* Changes with *Temperature*

1) The Q_{10} value for a reaction shows how much the <u>rate changes</u> when the <u>temperature</u> is <u>raised</u> by <u>10 °C</u>.

2) You can <u>calculate it</u> using this <u>equation</u>:

$$Q_{10} = \frac{\text{rate at higher temperature}}{\text{rate at lower temperature}}$$

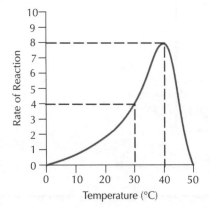

Temperature (°C)

3) The graph on the left shows the <u>rate of a reaction</u> between 0 °C and 50 °C. Here's how to calculate the Q_{10} value of the reaction using the rate at <u>30 °C</u> and at <u>40 °C</u>:

$$Q_{10} = \frac{\text{rate at 40 °C}}{\text{rate at 30 °C}} = \frac{8}{4} = 2$$

4) A Q_{10} value of <u>2</u> means that the <u>rate doubles</u> when the temperature is raised by 10 °C. A Q_{10} value of <u>3</u> would mean that the <u>rate trebles</u>.

Enzymes and Digestion

Not all enzymes work inside body cells — some work <u>outside</u> cells. For example, the enzymes used in <u>digestion</u> are produced by cells and then <u>released</u> into the <u>gut</u> to <u>mix</u> with <u>food</u>.

Digestive Enzymes Break Down Big Molecules into Smaller Ones

1) <u>Starch</u>, <u>proteins</u> and <u>fats</u> are big molecules.
   They're too big to pass through the walls of the digestive system.

2) <u>Sugars</u>, <u>amino acids</u>, <u>glycerol</u> and <u>fatty acids</u> are much smaller molecules.
   They can pass easily through the walls of the digestive system.

3) The <u>digestive enzymes</u> break down the big molecules into the smaller ones.

Amylase Converts Starch into Sugars

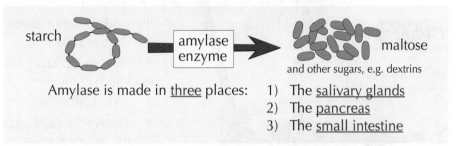

Amylase is made in <u>three</u> places: 1) The <u>salivary glands</u>
                                        2) The <u>pancreas</u>
                                        3) The <u>small intestine</u>

Protease Converts Proteins into Amino Acids

Protease is made in <u>three</u> places: 1) The <u>stomach</u> (it's called <u>pepsin</u> there)
                                         2) The <u>pancreas</u>
                                         3) The <u>small intestine</u>

Lipase Converts Lipids into Glycerol and Fatty Acids

Lipids are fats and oils.

Lipase is made in <u>two</u> places: 1) The <u>pancreas</u>
                                     2) The <u>small intestine</u>

Bile Neutralises the Stomach Acid and Emulsifies Fats

1) Bile is <u>produced</u> in the <u>liver</u>. It's <u>stored</u> in the <u>gall bladder</u> before it's released into the <u>small intestine</u>.

2) The <u>hydrochloric acid</u> in the stomach makes the pH <u>too acidic</u> for enzymes in the small intestine to work properly. Bile is <u>alkaline</u> — it <u>neutralises</u> the acid and makes conditions <u>alkaline</u>.
   The enzymes in the small intestine <u>work best</u> in these alkaline conditions.

3) It <u>emulsifies</u> fats. In other words it breaks the fat into <u>tiny droplets</u>. This gives a much <u>bigger surface area</u> of fat for the enzyme lipase to work on — which makes its digestion <u>faster</u>.

Enzymes and Digestion

So now you know what the enzymes do, here's a nice <u>big picture</u> of the <u>whole</u> of the digestive system.

The **Breakdown** of Food is Catalysed by **Enzymes**

1) Enzymes used in the digestive system are produced by specialised cells in <u>glands</u> and in the <u>gut lining</u>.

2) Different enzymes catalyse the <u>breakdown</u> of different food molecules.

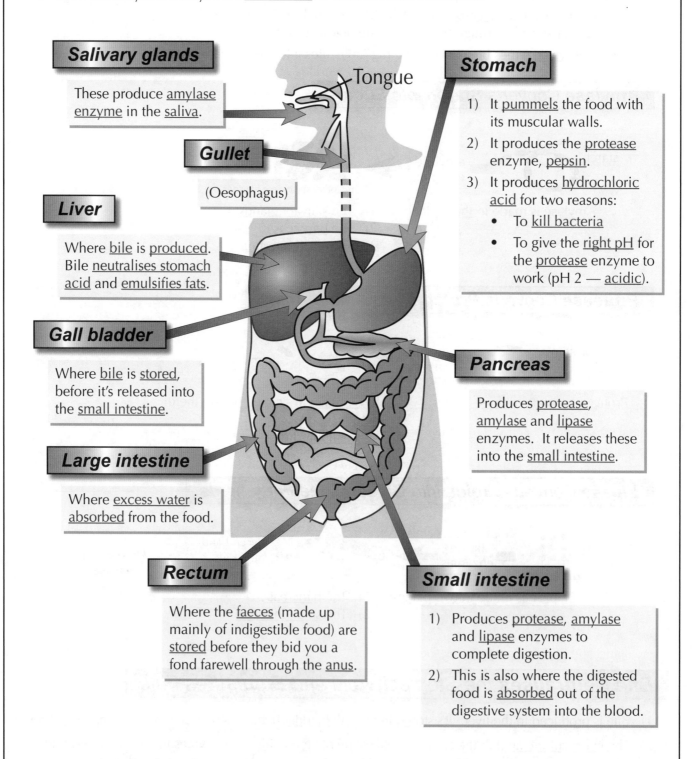

Salivary glands

These produce <u>amylase enzyme</u> in the <u>saliva</u>.

Tongue

Gullet

(Oesophagus)

Stomach

1) It <u>pummels</u> the food with its muscular walls.
2) It produces the <u>protease</u> enzyme, <u>pepsin</u>.
3) It produces <u>hydrochloric acid</u> for two reasons:
   - To <u>kill bacteria</u>
   - To give the <u>right pH</u> for the <u>protease</u> enzyme to work (pH 2 — <u>acidic</u>).

Liver

Where <u>bile</u> is <u>produced</u>. Bile <u>neutralises stomach acid</u> and <u>emulsifies fats</u>.

Gall bladder

Where <u>bile</u> is <u>stored</u>, before it's released into the <u>small intestine</u>.

Pancreas

Produces <u>protease</u>, <u>amylase</u> and <u>lipase</u> enzymes. It releases these into the <u>small intestine</u>.

Large intestine

Where <u>excess water</u> is <u>absorbed</u> from the food.

Rectum

Where the <u>faeces</u> (made up mainly of indigestible food) are <u>stored</u> before they bid you a fond farewell through the <u>anus</u>.

Small intestine

1) Produces <u>protease</u>, <u>amylase</u> and <u>lipase</u> enzymes to complete digestion.
2) This is also where the digested food is <u>absorbed</u> out of the digestive system into the blood.

There are nine different bits that make up the digestive system

Did you know that the whole of your digestive system is actually a big hole that goes right through your body? It just gets loads of food, digestive juices and enzymes piled into it...

Uses of Enzymes

Some <u>microorganisms</u> produce enzymes which pass <u>out</u> of their cells and catalyse reactions outside them (e.g. to <u>digest</u> the microorganism's <u>food</u>). These enzymes have many <u>uses</u> in the <u>home</u> and in <u>industry</u>.

Enzymes Are Used in **Biological Detergents**

1) <u>Enzymes</u> are the '<u>biological</u>' ingredients in biological detergents and washing powders.

2) They're mainly <u>protein-digesting</u> enzymes (<u>proteases</u>) and <u>fat-digesting</u> enzymes (<u>lipases</u>).

3) Because the enzymes break down <u>animal</u> and <u>plant</u> matter, they're ideal for removing <u>stains</u> like <u>food</u> or <u>blood</u>.

4) Biological detergents are also <u>more effective</u> at working at <u>low temperatures</u> (e.g. 30 °C) than other types of detergents.

Enzymes Are Used to **Change Foods**

1) The <u>proteins</u> in some <u>baby foods</u> are '<u>pre-digested</u>' using protein-digesting enzymes (<u>proteases</u>), so they're easier for the baby to digest.

2) Carbohydrate-digesting enzymes (<u>carbohydrases</u>) can be used to turn <u>starch syrup</u> into <u>sugar syrup</u>.

3) <u>Glucose syrup</u> can be turned into <u>fructose syrup</u> using an <u>isomerase</u> enzyme. Fructose is <u>sweeter</u>, so you can use <u>less</u> of it — good for slimming foods and drinks.

Using Enzymes in **Industry** Takes a Lot of **Control**

Enzymes are <u>really useful</u> in industry. They <u>speed up</u> reactions without the need for <u>high temperatures</u> and <u>pressures</u>. There are some <u>advantages</u> and <u>disadvantages</u> to using them, so here are a few to get you started:

ADVANTAGES:

1) They're <u>specific</u>, so they only catalyse the <u>reaction</u> you <u>want</u> them to.

2) Using lower temperatures and pressures means a <u>lower cost</u> as it <u>saves energy</u>.

3) Enzymes work for a <u>long time</u>, so after the <u>initial cost</u> of buying them, you can <u>continually</u> use them.

4) They are <u>biodegradable</u> and therefore cause less <u>environmental pollution</u>.

DISADVANTAGES:

1) Some people can develop <u>allergies</u> to the enzymes (e.g. in biological washing powders).

2) Enzymes can be <u>denatured</u> by even a <u>small</u> increase in temperature. They're also susceptible to <u>poisons</u> and changes in <u>pH</u>. This means the conditions in which they work must be <u>tightly controlled</u>.

3) Enzymes can be <u>expensive</u> to produce.

4) <u>Contamination</u> of the enzyme with other substances can affect the reaction.

From baby food to washing powder — enzymes make life easier

There's no denying that <u>enzymes</u> are <u>useful</u> and they have loads of <u>advantages</u>, but they're also <u>picky</u> — e.g. tiny changes in pH can stop them working. So the disadvantages are worth bearing in mind.

Warm-Up and Exam Questions

Doing well in exams isn't just about remembering all the facts, although that's important. You have to get used to the way the exams are phrased and make sure you always read the question carefully.

Warm-Up Questions

1) Enzymes are sometimes referred to as 'biological catalysts'. What is a catalyst?
2) What is meant by the optimum pH of an enzyme?
3) What is the function of digestive enzymes?
4) Which enzyme digests: (a) protein   (b) lipids?
5) What are the products of the digestion of: (a) starch   (b) protein   (c) lipids?
6) Explain why proteases are used in some baby food.

Exam Questions

1   The diagram represents the action of an enzyme in catalysing a biological reaction.

In terms of the enzyme's shape, explain the following:

(a)   why an enzyme only catalyses one reaction.

(1 mark)

(b)   what happens when the enzyme is denatured.

(1 mark)

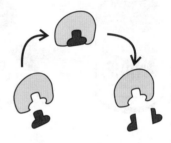

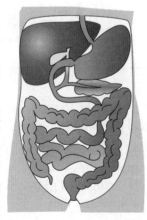

2   The diagram on the left shows the human digestive system.
    (a)   Copy the diagram and label the following parts:
       (i)   a part which is very acidic

(1 mark)

       (ii)   the place where bile is produced

(1 mark)

    (b)   Describe the functions of each of these parts of the digestive system:
       (i)   gall bladder

(1 mark)

       (ii)   pancreas

(2 marks)

3   Enzymes are used in the food industry to turn glucose syrup into fructose syrup.
    (a)   Name the enzyme used to turn glucose syrup into fructose syrup.

(1 mark)

    (b)   Explain why this process is carried out.

(2 marks)

Exam Questions

4 The graph below shows the effect of temperature on the action of two different enzymes.

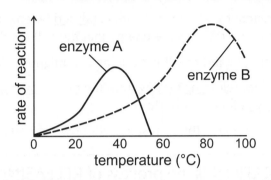

(a) What is the optimum temperature for enzyme **A**?

(1 mark)

(b) One of these enzymes was extracted from human liver cells.
The other was extracted from bacteria living in hot underwater vents.
Suggest which enzyme came from the bacteria. Give a reason for your answer.

(1 mark)

(c) Enzyme B is a protein-digesting enzyme.
Suggest why it might be useful in biological washing powders.

(2 marks)

5 Naz did an experiment to investigate the effect of pH on enzyme action. She took four
test tubes and placed some starch solution in each one. Each solution was given a
different pH value as shown below. Then a digestive enzyme was added to each tube.

Test tube	A	B	C	D
pH	2	5	8	11

(a) What type of digestive enzyme do you think should be added to the tubes?

(1 mark)

(b) Name three factors that should be kept constant during this experiment.

(3 marks)

(c) At which pH do you think the enzyme will work best? Explain your answer.

(2 marks)

6 *In this question you will be assessed on the quality of your English,
the organisation of your ideas and your use of appropriate specialist vocabulary.*

Discuss the advantages and disadvantages of using enzymes in industry.

(6 marks)

Respiration

Many chemical reactions inside cells are controlled by enzymes — including the ones in respiration.

Respiration is NOT "Breathing In and Out"

Respiration involves many reactions, all of which are catalysed by enzymes. These are really important reactions, as respiration releases the energy that the cell needs to do just about everything.

1) Respiration is not breathing in and breathing out, as you might think.

2) Respiration is the process of releasing energy from the breakdown of glucose — and it goes on in every cell in your body.

3) It happens in plants too. All living things respire. It's how they release energy from their food.

> RESPIRATION is the process of RELEASING ENERGY
> FROM GLUCOSE, which goes on IN EVERY CELL.

Aerobic Respiration Needs Plenty of Oxygen

1) Aerobic respiration is respiration using oxygen. It's the most efficient way to release energy from glucose. (You can also have anaerobic respiration, which happens without oxygen, but that doesn't release nearly as much energy — see page 30.)

2) Aerobic respiration goes on all the time in plants and animals.

3) Most of the reactions in aerobic respiration happen inside mitochondria (see page 13).

4) The overall word equation and the symbol equation for aerobic respiration are shown below:

Glucose + Oxygen $\Longrightarrow$ Carbon dioxide + Water (+ Energy Released)

$C_6H_{12}O_6 + 6O_2 \Longrightarrow 6CO_2 + 6H_2O$ (+ Energy Released)

reactants   products

Reactants are turned into products during a reaction.

Respiration Releases Energy for All Kinds of Things

Here are four examples of what the energy released by aerobic respiration is used for:

1) To build up larger molecules from smaller ones (like proteins from amino acids).

2) In animals, to allow the muscles to contract (which in turn allows them to move about).

3) In mammals and birds the energy is used to keep their body temperature steady (unlike other animals, mammals and birds keep their bodies constantly warm).

4) In plants, to build sugars, nitrates and other nutrients into amino acids, which are then built up into proteins.

Exercise

When you exercise, your body quickly adapts so that your muscles get more oxygen and glucose to supply energy. If your body can't get enough oxygen or glucose to them, it has some back-up plans ready.

Exercise *Increases* the *Heart Rate*

1) Muscles are made of muscle cells. These use oxygen to release energy from glucose (aerobic respiration — see page 28), which is used to contract the muscles.

2) An increase in muscle activity requires more glucose and oxygen to be supplied to the muscle cells. Extra carbon dioxide needs to be removed from the muscle cells. For this to happen the blood has to flow at a faster rate.

3) This is why physical activity:

- increases your breathing rate and makes you breathe more deeply to meet the demand for extra oxygen.

- increases the speed at which the heart pumps.

An unfit person's heart rate goes up a lot more during exercise than a fit person's, and they take longer to recover.

Glycogen is Used During Exercise

1) Some glucose from food is stored as glycogen.

2) Glycogen's mainly stored in the liver, but each muscle also has its own store.

3) During vigorous exercise muscles use glucose rapidly, so some of the stored glycogen is converted back to glucose to provide more energy.

Glucose is stored as glycogen in the liver and muscles

I bet you're exhausted after reading this page. The body really is a clever thing — it goes through all sorts of changes when you exercise — your breathing rate increases, your breathing depth increases and your heart rate increases too. All this helps plenty of glucose and oxygen to get to your muscles, and carbon dioxide to be taken away, which is just what you need to keep them working.

Exercise and Anaerobic Respiration

If your body can't get enough oxygen or glucose to your muscles, it has a back-up plan ready...

Anaerobic Respiration is Used if There's Not Enough Oxygen

1) When you do vigorous exercise and your body can't supply enough oxygen to your muscles, they start doing anaerobic respiration as well as aerobic respiration.

2) "Anaerobic" just means "without oxygen".
It's the incomplete breakdown of glucose, which produces lactic acid.

$$glucose \rightarrow energy + lactic\ acid$$

3) This is NOT the best way to convert glucose into energy because lactic acid builds up in the muscles, which gets painful. It also causes muscle fatigue — the muscles get tired and the stop contracting efficiently.

4) Another downside is that anaerobic respiration does not release nearly as much energy as aerobic respiration — but it's useful in emergencies.

5) The advantage is that at least you can keep on using your muscles for a while longer.

Anaerobic Respiration Leads to an Oxygen Debt

1) After resorting to anaerobic respiration, when you stop exercising you'll have an "oxygen debt".

2) In other words you have to "repay" the oxygen that you didn't get to your muscles in time, because your lungs, heart and blood couldn't keep up with the demand earlier on.

3) This means you have to keep breathing hard for a while after you stop, to get more oxygen into your blood. Blood flows through your muscles to remove the lactic acid by oxidising it to harmless CO_2 and water.

These rowers have finished rowing, but they're still breathing hard to replace their oxygen debt.

4) While high levels of CO_2 and lactic acid are detected in the blood (by the brain), the pulse and breathing rate stay high to try and rectify the situation.

Oxygen debt needs to be repaid

Yeast also respire anaerobically, but they produce ethanol (and carbon dioxide). So perhaps it's just as well humans produce lactic acid instead — or after a bit of vigorous exercise we'd all be drunk.

Diffusion

Particles <u>move about randomly</u>, and after a bit they end up <u>evenly spaced</u>.

Don't Be Put Off by the *Fancy Word*

1) "<u>Diffusion</u>" is simple. It's just the <u>gradual movement</u> of particles from places where there are <u>lots</u> of them to places where there are <u>fewer</u> of them.

2) That's all it is — just the <u>natural tendency</u> for stuff to <u>spread out</u>.

3) There's a fancy way of saying the same thing, which is this:

> <u>DIFFUSION</u> is the <u>spreading out</u> of <u>particles</u> from an area of <u>HIGH CONCENTRATION</u> to an area of <u>LOW CONCENTRATION</u>

4) Diffusion happens in both <u>liquids</u> and <u>gases</u> — that's because the <u>individual particles</u> in these substances are free to <u>move about</u> randomly.

5) The <u>simplest type</u> is when different <u>gases</u> diffuse through each other.
This is what's happening when the smell of perfume diffuses through the air in a room:

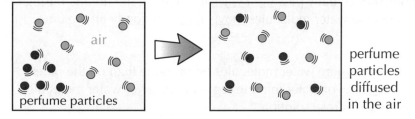

perfume particles diffused in the air

Cell Membranes are Kind of *Clever*

They're clever because they <u>hold</u> the cell together <u>BUT</u> they let stuff <u>in and out</u> as well. Only very <u>small molecules</u> can <u>diffuse</u> through cell membranes though — things like <u>simple sugars</u>, <u>water</u> or <u>ions</u>. <u>Big</u> molecules like <u>starch</u> and <u>proteins</u> can't pass through the membrane.

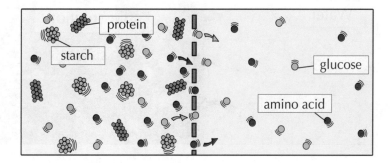

1) Just like with diffusion in air, particles flow through the cell membrane from where there's a <u>higher concentration</u> (more of them) to where there's a <u>lower concentration</u> (not such a lot of them).

2) They're only moving about <u>randomly</u> of course, so they go <u>both</u> ways — but if there are a lot <u>more</u> particles on one side of the membrane, there's a <u>net</u> (overall) movement <u>from</u> that side.

3) The <u>rate</u> of diffusion depends on three main things:

   a) <u>Distance</u> — substances diffuse <u>more quickly</u> when they haven't as <u>far</u> to move.

   b) <u>Concentration difference</u> (gradient) — substances diffuse faster if there's a <u>big difference</u> in concentration. If there are <u>lots more</u> particles on one side, there are more there to move across.

   c) <u>Surface area</u> — the <u>more surface</u> there is available for molecules to move across, the <u>faster</u> they can get from one side to the other.

Osmosis

If you've got your head round <u>diffusion</u>, osmosis will be a <u>breeze</u>. If not, have another look at page 31.

Osmosis is a Special Case of Diffusion, That's All

<u>OSMOSIS</u> is the <u>net movement of water molecules</u> across a <u>partially permeable membrane</u> from a region of <u>high water concentration</u> (i.e. a dilute solution) to a region of <u>low water concentration</u> (i.e. a concentrated solution).

1) A <u>partially permeable</u> membrane is just one with very small holes in it. So small, in fact, only tiny <u>molecules</u> (like water) can pass through them, and bigger molecules (e.g. <u>sucrose</u>) can't.

2) The water molecules actually pass <u>both ways</u> through the membrane during osmosis. This happens because water molecules <u>move about randomly</u> all the time.

3) But because there are <u>more</u> water molecules on one side than on the other, there's a steady <u>net flow</u> of water into the region with <u>fewer</u> water molecules, i.e. into the <u>stronger</u> sucrose solution.

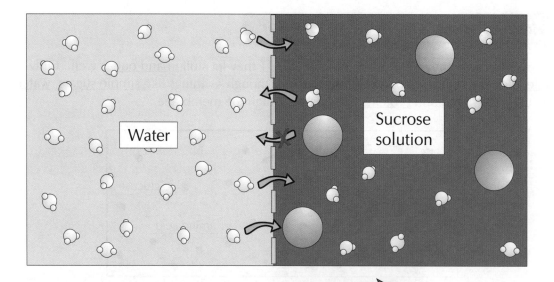

Net movement of water molecules

4) This means the <u>concentrated sucrose</u> solution gets more <u>dilute</u>. The water acts like it's trying to "<u>even up</u>" the concentration either side of the membrane.

5) Osmosis is a type of <u>diffusion</u> — net movement of <u>particles</u> from an area of <u>higher concentration</u> to an area of <u>lower concentration</u>.

Osmosis in Cells

The amount of water in cells has to be <u>just right</u> — too much or too little can mean problems...

Turgor Pressure Supports Plant Tissues

Normal Cell   Turgid Cell

1) When a plant is well watered, all its cells will draw water in by <u>osmosis</u> and become plump and swollen. When the cells are like this, they're said to be <u>turgid</u>.

2) The contents of the cell push against the <u>inelastic cell wall</u> — this is called <u>turgor pressure</u>. Turgor pressure helps <u>support</u> the plant tissues.

3) If there's no water in the soil, a plant starts to <u>wilt</u> (droop). This is because the cells start to lose water and so <u>lose</u> their turgor pressure. They're then said to be <u>flaccid</u>.

4) If the plant's really short of water, the <u>cytoplasm</u> inside its cells starts to <u>shrink</u> and the membrane <u>pulls away</u> from the cell wall. The cell is now said to be <u>plasmolysed</u>.

5) The plant doesn't totally lose its shape though, because the <u>inelastic cell wall</u> keeps things in position. It just droops a bit.

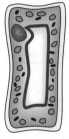
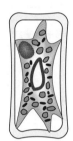

Flaccid Cell   Plasmolysed Cell

Animal Cells Don't Have an Inelastic Cell Wall

1) Plant cells aren't too bothered by changes in the amount of water because the <u>inelastic cell wall</u> keeps everything in place.

2) It's different in <u>animal cells</u> because they don't have a cell wall. If an animal cell <u>takes in</u> too much water, it <u>bursts</u> — this is known as <u>lysis</u>. If it <u>loses</u> too much water it gets all <u>shrivelled up</u> — this is known as <u>crenation</u>.

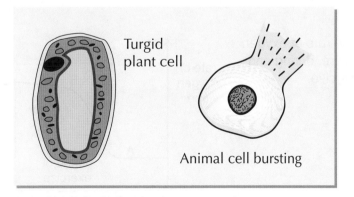

Turgid plant cell

Animal cell bursting

3) What all this means is that animals have to keep the amount of water in their cells pretty <u>constant</u> or they're in trouble, while plants are a bit more <u>tolerant</u> of periods of drought.

Warm-Up and Exam Questions

These questions might not be the most fun things in the world, but it's important you don't skip over them. They'll help you work out what you do and don't know, and what bits you need to read again.

Warm-Up Questions

1) Define respiration.
2) What substance, stored in the liver and muscles, is broken down during exercise to release glucose?
3) Define diffusion.
4) Describe a plasmolysed plant cell.

Exam Questions

1   Below are three diagrams showing cells surrounded by glucose.

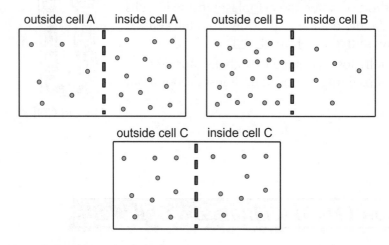

outside cell A   inside cell A   outside cell B   inside cell B

outside cell C   inside cell C

(a)   Into which cell, **A**, **B** or **C**, will there be a net movement of glucose? Explain your answer.

(2 marks)

(b)   Other than glucose, name **one** molecule that can diffuse through cell membranes into cells.

(1 mark)

2   The graph shows the rate of oxygen use by a person before, during and after a period of exercise.

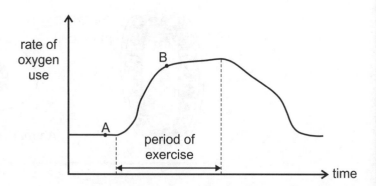

rate of oxygen use

B

A

period of exercise

time

(a)   Why is the rate of oxygen consumption higher at **B** than at **A**?

(2 marks)

(b)   Suggest why oxygen use remains high, even after the period of exercise ends.

(1 mark)

Blood Vessels

Blood is pumped around the body in different types of <u>blood vessels</u>.

Blood Vessels are Designed for Their Function

There are <u>three</u> different types of <u>blood vessel</u>:

> 1) <u>ARTERIES</u> — these carry the blood <u>away</u> from the heart.

> 2) <u>CAPILLARIES</u> — these are involved in the <u>exchange of materials</u> at the tissues.

> 3) <u>VEINS</u> — these carry the blood <u>to</u> the heart.

Arteries Carry Blood Under Pressure

> 1) The heart pumps the blood out at <u>high pressure</u> so the artery walls are <u>strong</u> and <u>elastic</u>.

> 2) The walls are <u>thick</u> compared to the size of the hole down the middle (the "<u>lumen</u>").

> 3) They contain thick layers of <u>muscle</u> to make them <u>strong</u>.

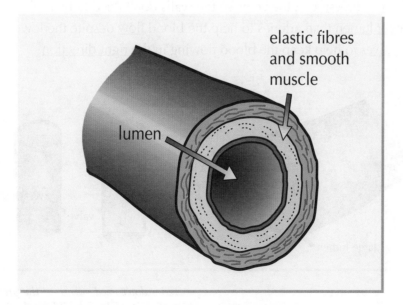

elastic fibres and smooth muscle

lumen

ARTERIES carry blood AWAY from the heart...

... an important fact there. Also, their <u>walls</u> are made of <u>thick muscle</u> and are <u>elastic</u> to withstand the <u>high pressure</u> of the <u>blood</u> being <u>pumped out</u> of the heart. In other words, they're adapted to their job...

Blood Vessels

This page is about the other two types of blood vessel — <u>capillaries</u> and <u>veins</u>.

Capillaries are Really Small

1) Arteries branch into <u>capillaries</u>.
2) Capillaries are really <u>tiny</u> — too small to see.
3) They carry the blood <u>really close</u> to <u>every cell</u> in the body to <u>exchange substances</u> with them.
4) They have <u>permeable</u> walls, so substances can <u>diffuse</u> in and out. (See page 31 for more on diffusion.)
5) They supply <u>food</u> and <u>oxygen</u>, and take away <u>waste</u> like CO_2.
6) Their walls are usually <u>only one cell thick</u>. This <u>increases</u> the rate of diffusion by <u>decreasing</u> the <u>distance</u> over which it occurs.

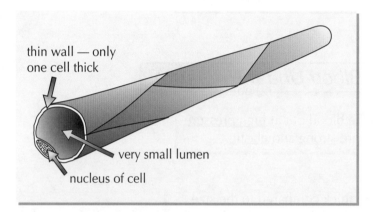

thin wall — only one cell thick

very small lumen

nucleus of cell

Veins Take Blood Back to the Heart

1) Capillaries eventually <u>join up</u> to form <u>veins</u>.
2) The blood is at <u>lower pressure</u> in the veins so the walls don't need to be as <u>thick</u> as artery walls.
3) They have a <u>bigger lumen</u> than arteries to help the blood <u>flow</u> despite the lower pressure.
4) They also have <u>valves</u> to help keep the blood flowing in the <u>right direction</u>.

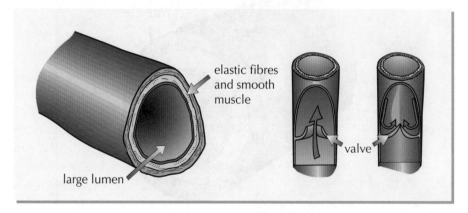

elastic fibres and smooth muscle

valve

large lumen

Arteries don't need valves — the pressure in them is high enough to keep the blood flowing the right way.

The blood vessels have different features for their different functions

Here's an interesting fact for you — your body contains about <u>60 000 miles</u> of blood vessels. That's about <u>six times</u> the distance from <u>London</u> to <u>Sydney</u> in Australia. Of course, capillaries are really tiny, which is how there can be such a big length — they can only be seen with a <u>microscope</u>.

The Heart

Blood doesn't just move around the body <u>on its own</u>, of course. It needs a <u>pump</u>.

Mammals Have a Double Circulatory System

1) The first system connects the <u>heart</u> to the <u>lungs</u>. <u>Deoxygenated</u> blood is pumped to the <u>lungs</u> to take in <u>oxygen</u>. The blood then <u>returns</u> to the heart.

2) The second system connects the <u>heart</u> to the <u>rest of the body</u>. The <u>oxygenated</u> blood in the heart is pumped out to the <u>body</u>. It <u>gives up</u> its oxygen, and then the <u>deoxygenated</u> blood <u>returns</u> to the heart to be pumped out to the <u>lungs</u> again.

3) Not all animals have a double circulatory system — <u>fish don't</u>, for example.

4) There are <u>advantages</u> to mammals having a double circulatory system though. Returning the blood to the <u>heart</u> after it's picked up oxygen at the <u>lungs</u> means it can be pumped out around the body at a much <u>higher pressure</u>. This <u>increases</u> the <u>rate of blood flow</u> to the tissues (i.e. blood can be pumped around the body much <u>faster</u>), so <u>more oxygen</u> can be delivered to the cells. This is important for mammals because they use up a lot of oxygen <u>maintaining their body temperature</u>.

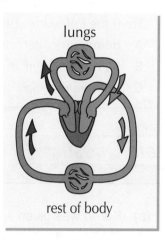

lungs

rest of body

The Heart has Four Chambers and Four Major Blood Vessels

Right Side   ### Left Side

(No, we haven't made a mistake — this is the right and left side of the person whose heart it is.)

pulmonary artery

vena cava

aorta

pulmonary vein

right atrium

left atrium

semi-lunar valve

semi-lunar valve

bicuspid valve

tricuspid valve

right ventricle

left ventricle

1) The <u>right atrium</u> of the heart receives <u>deoxygenated</u> blood from the <u>body</u> (through the <u>vena cava</u>). (The plural of atrium is atria.)

2) The deoxygenated blood moves through to the <u>right ventricle</u>, which pumps it to the <u>lungs</u> (via the <u>pulmonary artery</u>).

3) The <u>left atrium</u> receives <u>oxygenated</u> blood from the <u>lungs</u> (through the <u>pulmonary vein</u>).

4) The oxygenated blood then moves through to the <u>left ventricle</u>, which pumps it out round the <u>whole body</u> (via the <u>aorta</u>).

5) The <u>left</u> ventricle has a much <u>thicker wall</u> than the <u>right</u> ventricle. It needs more <u>muscle</u> because it has to pump blood around the <u>whole body</u>, whereas the right ventricle only has to pump it to the <u>lungs</u>.

6) The <u>semilunar</u>, <u>tricuspid</u> and <u>bicuspid valves</u> prevent the <u>backflow</u> of blood.

Warm-Up and Exam Questions

Hopefully I've persuaded you by now that it's a good idea to try these questions. So off you go...

Warm-Up Questions

1) Name the three types of blood vessel.
2) How many cells thick is the wall of a capillary?
3) From the following options, which one correctly describes the flow of blood through the heart?
   A   pulmonary artery and aorta, atria, ventricles, pulmonary vein and vena cava
   B   pulmonary vein and vena cava, ventricles, atria, pulmonary artery and aorta
   C   pulmonary vein and vena cava, atria, ventricles, pulmonary artery and aorta

Exam Questions

1   (a)   Karen was given a sheep's heart to look at. She made a drawing of the outside
         of the heart, and wrote some observations.

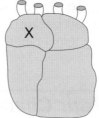

On the outside the heart is mainly
red in colour, with four big tubes
coming out of the top. The top part
of the heart feels quite soft, but at
the bottom it feels much firmer.

   (i)   List the names of the four big tubes.

(4 marks)

   (ii)   Name the part of the heart labelled **X**.

(1 mark)

   (b)   Karen then dissected the heart. Inside she found the heart chambers,
         and some flaps between the chambers.
   (i)   How many heart chambers would you expect her to find?

(1 mark)

   (ii)   What were the flaps she found between the chambers?

(1 mark)

   (iii)   What is the function of the flaps?

(1 mark)

2   The table below describes the functions of different parts of the circulatory system.
    Match the words **A**, **B**, **C** and **D** with the numbers **1 - 4** in the table.

A   Capillaries
B   Heart
C   Arteries
D   Veins

Structure	Function
1	Pumps blood around the body
2	Returns blood to the heart
3	Takes blood away from the heart
4	Carries blood to every cell in the body

(4 marks)

Revision Summary for Section 1

And where do you think you're going? It's no use just reading through and thinking you've got it all —
this stuff will only stick in your head if you've learnt it <u>properly</u>. And that's what these questions are for.

I won't pretend they'll be easy — they're not meant to be, but all the information's in the section
somewhere. Have a go at all the questions, then if there are any you can't answer, go back, look stuff up
and try again. Enjoy...

1) Name five parts of a cell that both plant and animal cells have.
2) Where is the DNA found in:
   a) bacterial cells,
   b) animal cells?
3) What is a tissue? What is an organ?
4) Give three examples of tissues in the human stomach, and say what job they do.
5) Name one organ system found in the human body.
6) What is an enzyme?
7) Describe the 'lock and key' model.
8) Name two things that affect how quickly an enzyme works.
9) What is a denatured enzyme?
10) What does a Q_{10} value show?
11) In which three places in the body is amylase produced?
12) Explain why the stomach produces hydrochloric acid.
13) Give two kinds of enzyme that would be useful in a biological washing powder.
14) Write the word equation for aerobic respiration.
15) Write the symbol equation for aerobic respiration.
16) Name three things that the energy released by respiration is used for.
17) What is anaerobic respiration? Give the word equation for anaerobic respiration in our bodies.
18) Explain how you repay an oxygen debt.
19) Name two substances that can't diffuse through cell membranes.
20) Explain what osmosis is.
21) What is turgor pressure?
22) Why do arteries need very muscular, elastic walls?
23) Explain how capillaries are adapted to their function.
24) Name the blood vessel that joins to the right ventricle of the heart. Where does it take the blood?
25) Why does the left ventricle have a thicker wall than the right ventricle?

DNA

Once people had found out that DNA was the molecule that carried the instructions for characteristics from your parents to you, scientists did loads of studies to try and work out its structure.

Chromosomes Are Really Long Molecules of DNA

1) DNA stands for deoxyribonucleic acid.

2) A DNA molecule has two strands coiled together in the shape of a double helix (two spirals), as shown in the diagram opposite.

3) Each of the two DNA strands is made up of lots of small groups called "nucleotides". Each nucleotide contains a small molecule called a "base".

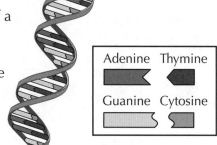

| Adenine | Thymine |
| Guanine | Cytosine |

4) The two strands are held together by these bases. There are four different bases (shown in the diagram as different colours) — adenine (A), cytosine (C), guanine (G) and thymine (T).

5) The bases are paired, and they always pair up in the same way — it's always A-T and C-G. This is called base-pairing (or complementary base-pairing).

6) The base pairs are joined together by weak hydrogen bonds.

Watson, Crick, Franklin and Wilkins Discovered DNA's Structure

1) Rosalind Franklin and Maurice Wilkins worked out that DNA had a helical structure by directing beams of X-rays onto crystallised DNA and looking at the patterns the X-rays formed as they bounced off.

2) James Watson and Francis Crick used these ideas, along with the knowledge that the amount of adenine + guanine matched the amount of thymine + cytosine, to make a model of the DNA molecule where all the pieces fitted together.

DNA Can Replicate Itself

1) DNA copies itself every time a cell divides, so that each new cell still has the full amount of DNA.

2) In order to copy itself, the DNA double helix first 'unzips' — to form two single strands.

3) New nucleotides (which float freely in the nucleus) then join on base-pairing (A with T and C with G). This makes an exact copy of the DNA on the other strand.

4) The result is two double-stranded molecules of DNA identical to the original molecule of DNA.

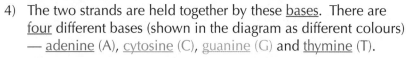

Molecule of DNA unzips.

Bases on free-floating nucleotides pair up with matching bases on the DNA.

Cross links form between the bases and the old DNA strands, and the nucleotides are joined together to form double strands.

Extracting DNA

If it's not good enough to just see a picture of DNA in a book, you can do an experiment that lets you see the real thing with your own eyes. That's what they call science in action...

You Can Do a **Practical** to **Extract DNA** From Cells

1) Chop up some <u>onion</u> and put it in a beaker containing a solution of <u>detergent</u> and <u>salt</u>. The detergent will <u>break down</u> the <u>cell membranes</u> and the salt will make the <u>DNA stick together</u>.

2) Put the beaker into a water bath at <u>60 °C</u> for <u>15 minutes</u> — this <u>denatures enzymes</u> (see p.21) that could digest the DNA and helps <u>soften</u> the onion cells.

3) Put the beaker in <u>ice</u> to <u>cool</u> the mixture down — this <u>stops</u> the DNA from <u>breaking down</u>.

4) Once the mixture is ice-cold, put it into a <u>blender</u> for a <u>few seconds</u> to <u>break open</u> the cell walls and <u>release</u> (but not break up) the DNA.

5) <u>Cool</u> the mixture down again, then <u>filter it</u> to get the froth and big bits of cell out.

6) <u>Gently</u> add some <u>ice-cold alcohol</u> to the filtered mixture. The <u>DNA</u> will start to <u>come out</u> of solution as it's <u>not soluble</u> in cold alcohol. It will appear as a <u>stringy white substance</u> that can be carefully fished out with a <u>glass rod</u>.

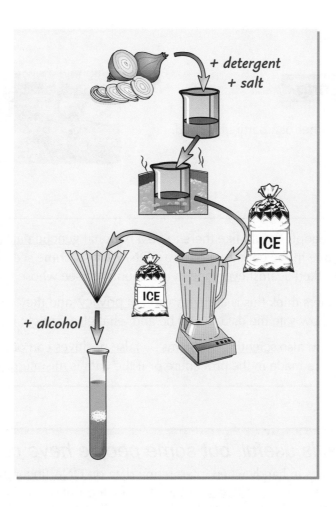

DNA Fingerprinting

Now this is interesting — you can use <u>DNA</u> to <u>catch criminals</u> or to <u>identify</u> the <u>father</u> of a child.

Everyone has *Unique* DNA...   *...except identical twins and clones*

1) Almost everyone's DNA is <u>unique</u>. The only exceptions are <u>identical twins</u>, where the two people have identical DNA, and <u>clones</u>.

2) <u>DNA fingerprinting</u> (or genetic fingerprinting) is a way of <u>cutting up</u> a person's DNA into small sections and then <u>separating</u> them.

3) Every person's genetic fingerprint has a <u>unique</u> pattern (unless they're identical twins or clones of course). This means you can <u>tell people apart</u> by <u>comparing samples</u> of their DNA.

DNA fingerprinting is used in...

1) Forensic Science

DNA (from hair, skin flakes, blood, semen etc.) taken from a <u>crime scene</u> is compared with a DNA sample taken from a suspect.

In the diagram, suspect 1's DNA has the same pattern as the DNA from the crime scene — so suspect 1 was probably at the crime scene.

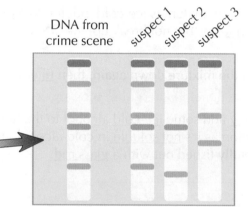

2) Paternity Testing

To see if a man is the father of a particular child.

- Some people would like there to be a national <u>genetic database</u> of everyone in the country. That way, DNA from a crime scene could be checked against <u>everyone</u> in the country to see whose it was.

- But others think this is a big <u>invasion of privacy</u>, and they worry about how <u>safe</u> the data would be and what <u>else</u> it might be used for.

- There are also <u>scientific problems</u> — <u>false positives</u> can occur if <u>errors</u> are made in the procedure or if the data is <u>misinterpreted</u>.

Forensic science is useful, but some people have concerns

Knowing about DNA comes in handy when interpreting data on <u>DNA fingerprinting for identification</u>. Say you're given the results of a <u>paternity test</u> — the DNA fingerprint of a child, their mother and some possible fathers. <u>Half</u> of the child's DNA fingerprint will <u>match</u> the <u>mother's DNA</u> fingerprint and <u>half</u> will match the <u>actual father's</u>.

Proteins

Your DNA is basically a long list of instructions on how to make <u>all the proteins</u> in your body.

A *Gene* Codes for a *Specific Protein*

1) A <u>gene</u> is a <u>section</u> of DNA. It contains the <u>instructions</u> to make a <u>specific protein</u>.
2) Cells make <u>proteins</u> by stringing <u>amino acids</u> together in a particular order.
3) Only <u>20</u> different amino acids are used to make up <u>thousands</u> of different <u>proteins</u>.
4) The <u>order of the bases</u> in a gene simply tells cells <u>in what order</u> to put the amino acids together:

> Each set of <u>three bases</u> (called a <u>triplet</u>) codes for a <u>particular amino acid</u>.
> Here's an <u>example</u> (don't worry — you don't have to remember the specific codes):
> TAT codes for tyrosine and GCA for alanine. If the order of the bases in the gene is
> TAT-GCA-TAT then the order of amino acids in the protein will be tyrosine-alanine-tyrosine.

5) DNA also determines which genes are <u>switched on or off</u> — and so which <u>proteins</u> the cell <u>produces</u>, e.g. haemoglobin or keratin. That in turn determines what <u>type of cell</u> it is, e.g. red blood cell, skin cell.
6) Some of the proteins <u>help to make</u> all the other things that <u>aren't made of protein</u> (like cell membranes) from substances that come from your diet (like fats and minerals).

Mutations can be *Harmful*, *Beneficial* or *Neutral*

A <u>mutation</u> is a <u>change</u> to an organism's <u>DNA base sequence</u>. This could affect the sequence of <u>amino acids</u> in the protein, which could affect the <u>shape</u> of the protein and so its <u>function</u>. In turn, this could affect the <u>characteristics</u> of an organism. Mutations can be <u>harmful</u>, <u>beneficial</u> or <u>neutral</u>:

Harmful

A mutation could cause a <u>genetic disorder</u>, for example <u>cystic fibrosis</u>.

Beneficial

A mutation could produce a <u>new characteristic</u> that is <u>beneficial</u> to an organism, e.g. mutations in genes on bacterial plasmids can make the bacteria <u>resistant</u> to <u>antibiotics</u>.

Neutral

Some mutations are <u>neither harmful nor beneficial</u>, e.g. they don't affect a protein's function.

Making Proteins

DNA controls which proteins are made, and when and where. But it can't do it alone.
Introducing <u>RNA</u> and the <u>ribosomes</u>...

Proteins are Made by Ribosomes

Proteins are made in the cell by <u>organelles</u> called <u>ribosomes</u>. DNA is found in the cell <u>nucleus</u> and
can't move out of it because it's <u>really big</u>. The cell needs to get the information from the DNA to the
<u>ribosome</u> in the cell cytoplasm. This is done using a molecule called <u>mRNA</u>, which is very similar
to DNA, but it's shorter and only a <u>single strand</u>. mRNA is like a <u>messenger</u> between the DNA in the
nucleus and the ribosome. Here's how it's done:

1) The two DNA strands <u>unzip</u>. The DNA is used as a
   <u>template</u> to make the <u>mRNA</u>. Base pairing ensures it's
   <u>complementary</u> (an exact match to the opposite strand).
   This step is called <u>TRANSCRIPTION</u>.

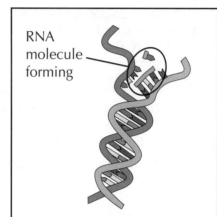

RNA
molecule
forming

2) The mRNA molecule <u>moves out</u> of the nucleus and <u>joins</u> with
   a ribosome.

3) <u>Amino acids</u> that match the mRNA code are
   <u>brought</u> to the ribosome by molecules called <u>tRNA</u>.

4) The job of the ribosome is to <u>stick amino acids</u>
   <u>together</u> in a chain to make a <u>polypeptide</u> (<u>protein</u>).
   This follows the order of the triplet of bases
   (called a <u>codon</u>) in the mRNA. This step is called <u>TRANSLATION</u>.

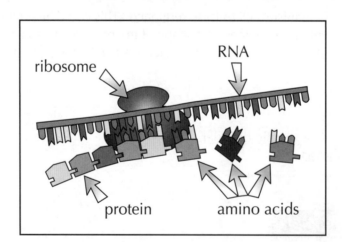

The result of all this molecular jiggery-pokery is that each type of <u>protein</u> gets made with its own specific
<u>number</u> and <u>sequence</u> of <u>amino acids</u> — the ones described by its <u>DNA base sequence</u>. This is what
makes it <u>fold up into the right shape</u> to do its specific <u>job</u>, e.g. as a particular <u>enzyme</u> (see page 20).

RNA carries the genetic code, ribosomes build the proteins

So, the order of <u>bases</u> in your DNA decides what <u>amino acids</u> get joined together, and the order of the
amino acids decides the type of <u>protein</u>. And proteins are pretty essential things — all your body's
enzymes (see page 20) are proteins, and enzymes control the making of your other, non-protein bits.

Warm-Up and Exam Questions

Question time again. Use these to find out what you know, and what you need to go back and read again. There's no point ignoring the questions you can't answer — make sure you're ready.

Warm-Up Questions

1) What does DNA stand for?
2) Name the four different bases in DNA and say how they pair up.
3) Explain why detergents and salt are used to extract DNA from cells.
4) Describe what a gene is.
5) Give an example of a beneficial mutation.

Exam Questions

1   Mr X and Mr Y are both suspects in a burglary. A blood stain has been found on a crowbar at the crime scene. The police carry out a DNA fingerprint on Mr X, Mr Y and the blood from the crime scene.

The diagram shows part of the test results.

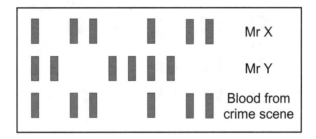

(a)   Do the results suggest that either of the suspects were at the crime scene? Explain your answer.

(1 mark)

(b)   A police officer investigating the burglary says that no two people have exactly the same genetic fingerprint. Is he correct? Explain your answer.

(2 marks)

2   Part of one of the strands of a DNA molecule has this base sequence:

C — C — G — T — T — T — G — G — G

(a)   What is the base sequence of the equivalent part of the other DNA strand?

(1 mark)

(b)   Describe how DNA replicates itself.

(3 marks)

(c)   Watson and Crick used data from other scientists to model the structure of DNA in 1953. Describe what evidence they used.

(2 marks)

Cell Division — Mitosis

In order to <u>survive</u> and <u>grow</u>, our cells have got to be able to <u>divide</u>. And that means our <u>DNA</u> as well...

Mitosis Makes New Cells For **Growth** and **Repair**

1) Human <u>body cells</u> are <u>diploid</u>. This means they normally have <u>two versions</u> of each <u>chromosome</u> — one from the person's <u>mother</u>, and one from their <u>father</u>.

2) The diagram below shows the <u>23 pairs of chromosomes</u> in a human cell.

3) When a cell <u>divides</u> it makes <u>two</u> cells <u>identical</u> to the <u>original</u> cell — each with a <u>nucleus</u> containing the <u>same number</u> of chromosomes as the original cell.

4) This type of cell division is called <u>mitosis</u>. It's used when humans (and animals and plants) want to <u>grow</u> or <u>replace</u> cells that have been <u>damaged</u>.

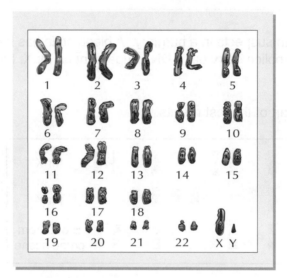

Asexual Reproduction Also Uses **Mitosis**

1) Some organisms also <u>reproduce</u> by mitosis, e.g. strawberry plants can form runners in this way, which become new plants.

2) This is an example of <u>asexual</u> reproduction.

3) The offspring have exactly the <u>same genes</u> as the parent — so there's <u>no genetic variation</u>.

Mitosis happens in organisms for growth and repair
The next page is all about how mitosis happens.

Cell Division — Mitosis

So, you can make new cells using mitosis. Here's how it actually happens...

Mitosis Results in Two Identical Cells

In a cell that's not dividing, the DNA is all spread out in <u>long strings</u>.

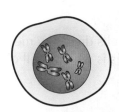

If the cell gets a signal to <u>divide</u>, it needs to <u>duplicate</u> its DNA — so there's one copy for each new cell. The DNA is copied and forms <u>X-shaped</u> chromosomes. Each 'arm' of the chromosome is an <u>exact duplicate</u> of the other.

The left arm of the chromosome has the same DNA as the right arm.

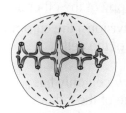

The chromosomes then <u>line up</u> at the centre of the cell and <u>cell fibres</u> pull them apart. The <u>two arms</u> of each chromosome go to <u>opposite ends</u> of the cell.

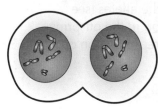

<u>Membranes</u> form around each of the sets of chromosomes. These become the <u>nuclei</u> of the two new cells.

Lastly, the <u>cytoplasm</u> divides.

You now have <u>two new diploid cells</u> containing exactly the same DNA — they're <u>genetically identical</u>.

Cancer cells have no limit on the number of times they divide

Mitosis produces identical cells, but there's another type which doesn't (see next page).

Cell Division — Meiosis

Gametes Have Half the Usual Number of Chromosomes

1) Gametes are 'sex cells'. They're called ova (single, ovum) in females, and sperm in males. During sexual reproduction, two gametes combine to form a new cell which will grow to become a new organism

2) Gametes are haploid — this means they only have one copy of each chromosome. This is so that when two gametes combine at fertilisation, the resulting cell (zygote) has the right number of chromosomes. Zygotes are diploid — they have two copies of each chromosome.

3) For example, human body cells have 46 chromosomes. The gametes have 23 chromosomes each, so that when an egg and sperm combine, you get 46 chromosomes again.

Meiosis Involves Two Divisions

To make new cells which only have half the original number of chromosomes, cells divide by meiosis. Meiosis only happens in the reproductive organs (i.e. ovaries and testes). Meiosis is when a cell divides to produce four haploid nuclei whose chromosomes are NOT identical.

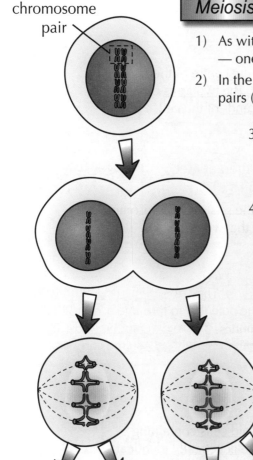

chromosome pair

Meiosis — Division 1

1) As with mitosis, before the cell starts to divide, it duplicates its DNA — one arm of each chromosome is an exact copy of the other arm.

2) In the first division in meiosis (there are two divisions) the chromosome pairs (see previous page) line up in the centre of the cell.

3) They're then pulled apart, so each new cell only has one copy of each chromosome. Some of the father's chromosomes (shown in blue) and some of the mother's chromosomes (shown in red) go into each new cell.

4) Each new cell will have a mixture of the mother's and father's chromosomes. Mixing up the alleles (see page 57) in this way creates variation in the offspring. This is a huge advantage of sexual reproduction over asexual reproduction.

Meiosis — Division 2

5) In the second division the chromosomes line up again in the centre of the cell. It's a lot like mitosis. The arms of the chromosomes are pulled apart.

6) You get four haploid gametes, each with only a single set of chromosomes in it.

After two gametes join at fertilisation, the cell grows by repeatedly dividing by mitosis.

Stem Cells

Most cells have specific features that make them particularly suited to the job that they do. But stem cells are a bit like a blank canvas — they have the potential to turn into other types of cells. And because of that, they're very important little cells...

Embryonic Stem Cells Can Turn into ANY Type of Cell

1) <u>Differentiation</u> is the process by which a cell <u>changes</u> to become <u>specialised</u> for its job (see pages 15-16).

2) In most <u>animal</u> cells, the ability to differentiate is <u>lost</u> at an early stage, but lots of <u>plant</u> cells <u>don't</u> ever lose this ability.

3) Some cells are <u>undifferentiated</u>. They can develop into <u>different types of cell</u> depending on what <u>instructions</u> they're given. These cells are called <u>STEM CELLS</u>.

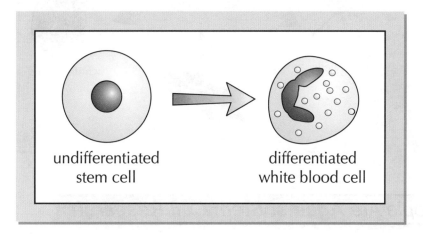

undifferentiated
stem cell

differentiated
white blood cell

4) Stem cells are found in early <u>human embryos</u>. They're <u>exciting</u> to doctors and medical researchers because they have the potential to turn into <u>any</u> kind of cell at all. This makes sense if you think about it — <u>all</u> the <u>different types</u> of cell found in a human being have to come from those <u>few cells</u> in the early embryo.

5) Adults also have stem cells, but they're only found in certain places, like <u>bone marrow</u>. These aren't as <u>versatile</u> as embryonic stem cells — they can't turn into <u>any</u> cell type at all, only certain ones.

Stem cells are pretty versatile little cells...

Your cells are pretty highly <u>specialised</u> for the jobs that they do — think how different a red blood cell is from any other cell. But you can trace <u>all</u> cells back to undifferentiated <u>stem cells</u> in the embryo.

Stem Cells

Stem cell research has exciting possibilities, but it's also pretty <u>controversial</u>.

Stem Cells May Be Able to Cure Many Diseases

1) Medicine <u>already</u> uses adult stem cells to cure <u>disease</u>. For example, people with some <u>blood diseases</u> (e.g. <u>sickle cell anaemia</u>) can be treated by <u>bone marrow transplants</u>. Bone marrow contains <u>stem cells</u> that can turn into <u>new blood cells</u> to replace the faulty old ones.

2) Scientists can also <u>extract</u> stem cells from very early human embryos and <u>grow</u> them.

3) These embryonic stem cells could be used to <u>replace faulty cells</u> in sick people — you could make <u>beating heart muscle cells</u> for people with <u>heart disease</u>, <u>insulin-producing cells</u> for people with <u>diabetes</u>, <u>nerve cells</u> for people <u>paralysed by spinal injuries</u>, and so on.

4) To get cultures of <u>one specific type</u> of cell, researchers try to <u>control</u> the differentiation of the stem cells by changing the environment they're growing in. So far, it's still a bit hit and miss — lots more <u>research</u> is needed.

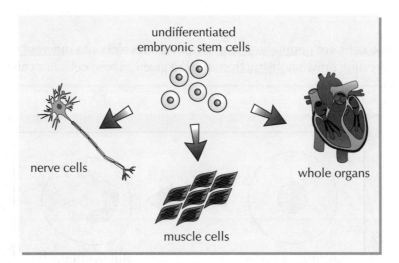

Some People Are Against Stem Cell Research

1) Some people are <u>against</u> stem cell research because they feel that human embryos <u>shouldn't</u> be used for experiments since each one is a <u>potential human life</u>.

2) Others think that curing patients who <u>already exist</u> and who are <u>suffering</u> is more important than the rights of <u>embryos</u>.

3) One fairly convincing argument in favour of this point of view is that the embryos used in the research are usually <u>unwanted ones</u> from <u>fertility clinics</u> that would probably just be <u>destroyed</u> if they weren't used for research. But of course, campaigners for the rights of embryos usually want this banned too.

4) These campaigners feel that scientists should concentrate more on finding and developing <u>other sources</u> of stem cells, so people could be helped <u>without</u> having to use embryos.

5) In some countries stem cell research is <u>banned</u>, but it's allowed in the UK as long as it follows <u>strict guidelines</u>.

Alternative sources of stem cells would avoid the controversy

Research has been done into getting stem cells from <u>other sources</u> — for example, some scientists think it might be possible to get cells from <u>umbilical cords</u> to behave like embryonic stem cells.

Growth and Development

<u>Growth</u> is an increase in <u>size</u> or <u>mass</u>. It can be <u>measured</u> in many different ways — read on...

Animals **Stop** Growing, Plants Can Grow **Continuously**

Plants and animals <u>grow differently</u>:

1) Animals tend to grow until they reach a <u>finite size</u> (full growth) and then <u>stop</u> growing. Plants often grow <u>continuously</u> — even really old trees will keep putting out <u>new branches</u>.

2) In animals, growth happens by <u>cell division</u>. In plants, growth in <u>height</u> is mainly due to <u>cell enlargement</u> (elongation). Growth by cell <u>division</u> usually just happens in areas of the plant called <u>meristems</u> (at the <u>tips</u> of the <u>roots</u> and <u>shoots</u>).

There are **Different Methods** for **Measuring Growth**

To work out if something's grown (i.e. increased in size), you need to take more than one measurement.

Growth of plants and animals can be quite <u>tricky</u> to <u>measure</u> — there are <u>different methods</u>, but they all have <u>pros</u> and <u>cons</u>.

Method	What it involves	Advantages	Disadvantages
LENGTH	Just measure the <u>length</u> (or <u>height</u>) of a plant or animal.	<u>Easy</u> to measure.	It <u>doesn't tell you</u> about changes in <u>width</u>, <u>diameter</u>, number of <u>branches</u>, etc.
WET MASS	<u>Weigh</u> the plant or animal and you have the wet mass.	<u>Easy</u> to measure.	Wet mass is very <u>changeable</u>. For example, a plant will be <u>heavier</u> if it's recently <u>rained</u> because it will have absorbed lots of water. Animals will be heavier if they've just <u>eaten</u> or if they've got a <u>full bladder</u>.
DRY MASS	<u>Dry out</u> the organism <u>before</u> <u>weighing</u> it.	It's <u>not affected</u> by the amount of <u>water</u> in a plant or animal or how much an organism has <u>eaten</u>.	You have to <u>kill</u> the organism to work it out. This might be okay for an area of <u>grass</u>, but it's <u>not so good</u> if you want to know the dry mass of a <u>person</u>.

<u>Dry mass</u> is actually the <u>best measure</u> of growth in plants and animals — it's <u>not affected</u> by changes in <u>water content</u> and it tells you the <u>size</u> of the <u>whole organism</u>.

One method might be more suitable than another...

... you just need to weigh up the advantages and disadvantages of each method and decide which one to use. Some methods might not be possible in some situations — for example, you can't go drying out every organism you want to measure, even if it is the best measure of growth.

Growth and Development

You can split human growth neatly up into <u>five different stages</u>. Growth happens at different rates in the different stages. And guess what... you can show this on a graph. Read on...

Human Growth has Different Phases

1) Humans go through <u>five main phases</u> of <u>growth</u> — as shown in the table below.

2) The <u>two main phases</u> of <u>rapid growth</u> take place <u>just after birth</u> and during <u>adolescence</u>. Growth stops when a person reaches adulthood.

Phase	Description
Infancy	Roughly the first two years of life. Rapid growth.
Childhood	Period between infancy and puberty. Steady growth.
Adolescence	Begins with puberty and continues until body development and growth are complete. Rapid growth.
Maturity/adulthood	Period between adolescence and old age. Growth stops.
Old age	Usually considered to be between age 65 and death.

3) The graph below is an example of a <u>typical human growth curve</u>. It shows how weight increases for boys between the ages of 2 and 20. When the line is <u>steeper</u>, growth is <u>more rapid</u> (e.g. during adolescence).

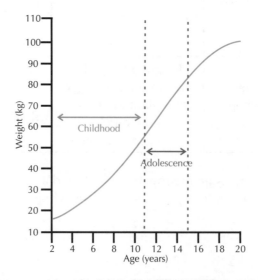

Growth happens at different rates from infancy to old age

Human growth curves are really good for showing how growth changes throughout life. You can use an average growth curve to compare someone's weight against the 'average' for their age.

Warm-Up and Exam Questions

There's only one way to do well in the exam — learn the facts and then practise lots of exam questions to see what it'll be like on the big day. We couldn't have made it easier for you — so do it.

Warm-Up Questions

1) What is mitosis?
2) In asexual reproduction, how many parents are there?
3) What type of cell division is involved in the regeneration of body cells?
4) Give one example of a disease that can be treated using stem cells.

Exam Questions

1 (a) What are stem cells?

(2 marks)

(b) Describe how stem cells could be used to treat disorders.

(1 mark)

(c) *In this question you will be assessed on the quality of your English, the organisation of your ideas and your use of appropriate specialist vocabulary.*

Discuss why some people are in favour of using embryos to create stem cells for research, while others are against the idea.

(6 marks)

2 (a) What are the male and female human gametes?

(1 mark)

(b) How many chromosomes does a human gamete contain?

(1 mark)

Two human gametes fuse to form a zygote.

(c) How many chromosomes does the zygote contain?

(1 mark)

(d) What fraction of its chromosomes has the zygote inherited from its mother?

(1 mark)

3 (a) The diagram below shows the chromosomes of a cell that is about to divide by meiosis.

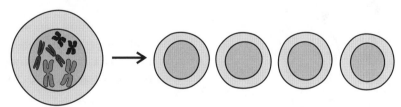

(i) Copy and complete the diagram to show the chromosomes in the new cells.

(2 marks)

(ii) How is the genetic content of the new cells different from the original cell?

(1 mark)

(b) State **three** ways in which meiosis is different from mitosis.

(3 marks)

X and Y Chromosomes

Now for a couple of very important little chromosomes...

Your **Chromosomes** Control Whether You're **Male** or **Female**

1) There are 22 matched pairs of chromosomes in every human body cell.

2) The 23rd pair are labelled XX or XY.

3) They're the two chromosomes that decide whether you turn out male or female.

> All men have an X and a Y chromosome: XY
> The Y chromosome causes male characteristics.

> All women have two X chromosomes: XX
> The XX combination allows
> female characteristics to develop.

X-chromosome   Y-chromosome

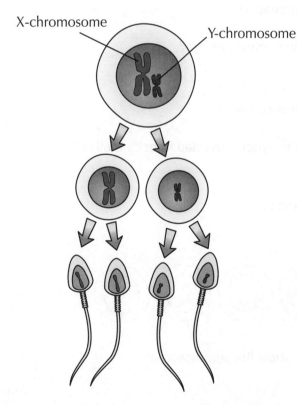

When making sperm, the X and Y chromosomes are drawn apart in the first division in meiosis (page 48). There's a 50% chance each sperm cell gets an X chromosome and a 50% chance it gets a Y chromosome.

A similar thing happens when making eggs. But the original cell has two X chromosomes, so all the eggs have one X chromosome.

The Y chromosome is physically smaller than the X chromosome

In some cases, it's possible for people to have one X and two Y chromosomes, or even three X chromosomes, in their cells. But usually, it's XX for girls and XY for boys.

X and Y Chromosomes

You can work out the <u>probability</u> of offspring being male or female by using a <u>genetic diagram</u>.

Genetic Diagrams Show the **Possible Combinations** of Gametes

1) To find the <u>probability</u> of getting a boy or a girl, you can draw a <u>genetic diagram</u>.

2) Put the <u>possible gametes</u> from <u>one</u> parent down the side, and those from the <u>other</u> parent along the top.

3) Then in each middle square you <u>fill in</u> the letters from the top and side that line up with that square. The <u>pairs of letters</u> in the middle show the possible combinations of the gametes.

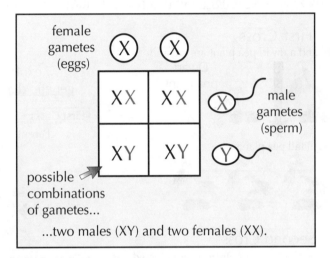

4) There are <u>two XX results</u> and <u>two XY results</u>, so there's the same probability of getting a boy or a girl.

5) Don't forget that this <u>50:50 ratio</u> is only a <u>probability</u> at each pregnancy. If you had four kids they <u>could</u> all be <u>boys</u>.

There's **More Than One Type** of Genetic Diagram

The other type of genetic diagram looks a bit more complicated, but it shows exactly the same thing.

1) At the top are the <u>parents</u>.

2) The middle circles show the <u>possible gametes</u> that are formed. One gamete from the female combines with one gamete from the male (during fertilisation).

3) The criss-cross lines show <u>all</u> the <u>possible</u> ways the X and Y chromosomes <u>could</u> combine.

4) The <u>possible combinations</u> of the offspring are shown in the bottom circles.

5) Remember, only <u>one</u> of these possibilities would <u>actually happen</u> for any one offspring.

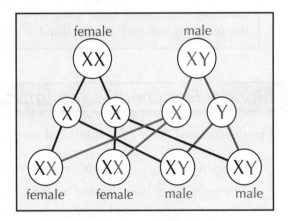

These diagrams aren't as scary as they look...

Most genetic diagrams you'll see concentrate on a <u>gene</u>, instead of a <u>chromosome</u>.
But the principle's the same. Don't worry — there are loads of other examples on the following pages.

The Work of Mendel

Gregor Mendel was pretty much the <u>founder of genetics</u>. Here's a whole page on him.

Mendel Did Genetic Experiments with Pea Plants

1) <u>Gregor Mendel</u> was an Austrian monk who trained in <u>mathematics</u> and <u>natural history</u> at the University of Vienna. On his garden plot at the monastery, Mendel noted how <u>characteristics</u> in <u>plants</u> were <u>passed on</u> from one generation to the next.

2) The results of his research were published in <u>1866</u> and eventually became the <u>foundation</u> of modern <u>genetics</u>.

3) These diagrams show two <u>crosses for height</u> in <u>pea plants</u> that Mendel carried out...

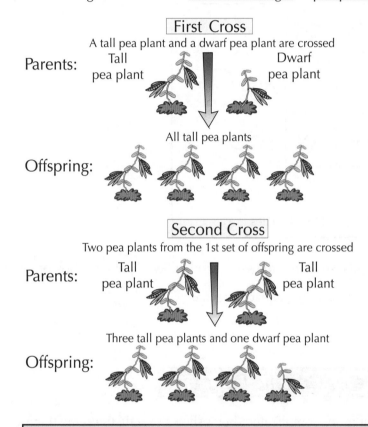

First Cross

A tall pea plant and a dwarf pea plant are crossed

Parents: Tall pea plant — Dwarf pea plant

All tall pea plants

Offspring:

Second Cross

Two pea plants from the 1st set of offspring are crossed

Parents: Tall pea plant — Tall pea plant

Three tall pea plants and one dwarf pea plant

Offspring:

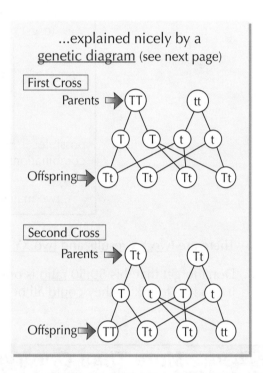

...explained nicely by a <u>genetic diagram</u> (see next page)

First Cross
Parents ➡ TT tt
T T t t
Offspring ➡ Tt Tt Tt Tt

Second Cross
Parents ➡ Tt Tt
T t T t
Offspring ➡ TT Tt Tt tt

Mendel had shown that the height characteristic in pea plants was determined by separately inherited "<u>hereditary units</u>" passed on from each parent. The ratios of tall and dwarf plants in the offspring showed that the unit for tall plants, <u>T</u>, was <u>dominant</u> over the unit for dwarf plants, <u>t</u>.

Mendel Reached Three Important Conclusions

Mendel reached these three important conclusions about <u>heredity in plants</u>:

1) Characteristics in plants are determined by "<u>hereditary units</u>".

2) Hereditary units are passed on from both parents, <u>one unit</u> from <u>each parent</u>.

3) Hereditary units can be <u>dominant</u> or <u>recessive</u> — if an individual has <u>both</u> the dominant and the recessive unit for a characteristic, the <u>dominant</u> characteristic will be expressed.

We now know that the "hereditary units" are, of course, <u>genes</u>.

But in Mendel's time <u>nobody</u> knew anything about genes or DNA, and so the <u>significance</u> of his work was not to be realised until <u>after his death</u>.

Genetic Diagrams

When a <u>single gene</u> controls the inheritance of a characteristic, you can work out the odds of getting it...

Genetic Diagrams Show the **Possible Genes** of Offspring

1) <u>Alleles</u> are <u>different versions</u> of the <u>same gene</u>.

2) In genetic diagrams <u>letters</u> are usually used to represent <u>alleles</u>.

Remember, gametes only have one allele, but all the other cells in an organism have two.

3) If an organism has <u>two alleles</u> for a particular gene <u>the same</u>, then it's <u>HOMOZYGOUS</u>. If its two alleles for a particular gene are <u>different</u>, then it's <u>HETEROZYGOUS</u>.

4) If the two alleles are <u>different</u>, only one can determine what <u>characteristic</u> is present. The allele for the <u>characteristic that's shown</u> is called the <u>dominant</u> allele (use a capital letter for dominant alleles — e.g. 'C'). The other one is called <u>recessive</u> (and you show these with small letters — e.g. 'c').

5) For an organism to display a <u>recessive</u> characteristic, <u>both</u> its alleles must be <u>recessive</u> (e.g. cc). But to display a <u>dominant</u> characteristic the organism can be <u>either</u> CC or Cc, because the dominant allele <u>overrules</u> the recessive one if the plant/animal/other organism is heterozygous.

It's Not Too Tricky to **Interpret**, **Explain** and **Construct** Them

Imagine you're cross-breeding <u>hamsters</u>, some with normal hair and a mild disposition and others with wild scratty hair and a leaning towards crazy acrobatics.

Let's say that the gene which causes the <u>crazy</u> nature is <u>recessive</u>, so we use a <u>small</u> "b" for it, whilst <u>normal</u> (boring) behaviour is due to a <u>dominant</u> gene, so we represent it with a <u>capital</u> "B".

1) A <u>crazy</u> hamster <u>must</u> have the <u>genotype bb</u>. However, a normal hamster could have <u>two</u> possible genotypes — BB or Bb.

<u>Genotype</u> means what alleles you have. <u>Phenotype</u> means the actual characteristic.

2) Here's what happens if you breed from two <u>homozygous</u> hamsters:

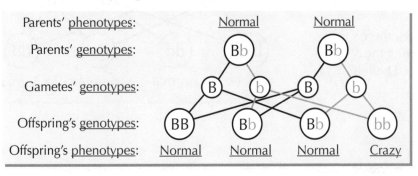

3) If two of these <u>offspring</u> now <u>breed</u>, you'll get the next generation:

When you cross two parents to look at just one characteristic, it's called a monohybrid cross.

Parents' phenotypes:	Normal		Normal	
Parents' genotypes:	Bb		Bb	
Gametes' genotypes:	B	b	B	b
Offspring's genotypes:	BB	Bb	Bb	bb
Offspring's phenotypes:	<u>Normal</u>	<u>Normal</u>	<u>Normal</u>	<u>Crazy</u>

4) This gives a <u>3:1 ratio</u> of normal to crazy offspring in this generation. Remember that "results" like this are only <u>probabilities</u> — they don't say definitely what'll happen.

Genetic Disorders

It's not just characteristics that are passed on — some <u>disorders</u> are inherited.

Cystic Fibrosis is Caused by a Recessive Allele

<u>Cystic fibrosis</u> is a <u>genetic disorder</u> of the <u>cell membranes</u>. It <u>results</u> in the body producing a lot of thick sticky <u>mucus</u> in the <u>air passages</u> and in the <u>pancreas</u>.

1) The allele which causes cystic fibrosis is a <u>recessive allele</u>, 'f', carried by about <u>1 person in 25</u>.

2) Because it's recessive, people with only <u>one copy</u> of the allele <u>won't</u> have the disorder — they're known as <u>carriers</u>.

3) For a child to have the disorder, <u>both parents</u> must be either <u>carriers</u> or <u>sufferers</u>.

4) There's a <u>1 in 4 chance</u> of a child having the disorder if <u>both</u> parents are <u>carriers</u>:

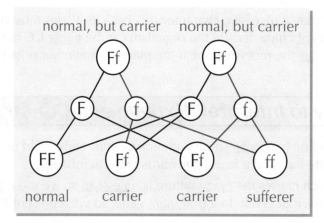

Polydactyly is Caused by a Dominant Allele

<u>Polydactyly</u> is a <u>genetic disorder</u> where a baby's born with <u>extra fingers or toes</u>. It doesn't usually cause any other problems so <u>isn't life-threatening</u>.

1) The disorder is caused by a <u>dominant allele</u>, 'D', and so can be inherited if just <u>one parent</u> carries the defective allele.

2) The <u>parent</u> that <u>has</u> the defective allele will be a <u>sufferer</u> too since the allele is dominant.

3) As the genetic diagram shows, there's a <u>50% chance</u> of a child having the disorder if <u>one</u> parent has the D allele.

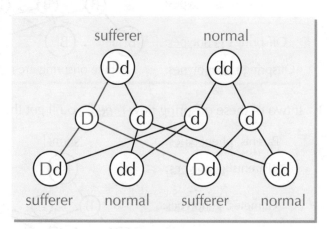

That's four genetic diagrams in two pages — ouch...

Genetic diagrams look pretty confusing at first, but they're really <u>not</u>. The important thing to get clear in your head is whether the <u>characteristic</u> is caused by a <u>dominant</u> or <u>recessive</u> allele.

Screening for Genetic Disorders

In vitro fertilisation (IVF) is quite widely used now by people who have problems conceiving naturally. Part of the process involves screening for genetic disorders, but some people are unhappy about this.

Embryos Can Be Screened for Genetic Disorders

1) During *in vitro* fertilisation (IVF), embryos are fertilised in a laboratory, and then implanted into the mother's womb. More than one egg is fertilised, so there's a better chance of the IVF being successful.

2) Before being implanted, it's possible to remove a cell from each embryo and analyse its genes.

3) Many genetic disorders could be detected in this way, such as cystic fibrosis.

4) Embryos with 'good' alleles would be implanted into the mother — the ones with 'bad' alleles destroyed.

There is a huge debate raging about embryonic screening. Here are some arguments for and against it.

Against Embryonic Screening

1) There may come a point where everyone wants to screen their embryos so they can pick the most 'desirable' one, e.g. they want a blue-eyed, blond-haired, intelligent boy.

2) The rejected embryos are destroyed — they could have developed into humans.

3) It implies that people with genetic problems are 'undesirable' — this could increase prejudice.

4) Screening is expensive.

For Embryonic Screening

1) It will help to stop people suffering.

2) There are laws to stop it going too far. At the moment parents cannot even select the sex of their baby (unless it's for health reasons).

3) During IVF, most of the embryos are destroyed anyway — screening just allows the selected one to be healthy.

4) Treating disorders costs the Government (and the taxpayers) a lot of money.

Many people think that embryonic screening isn't justified for genetic disorders that don't affect a person's health, such as polydactyly (see page 58).

Embryonic screening — it's a tricky one...

Like many things in science, there's a lot of debate surrounding embryonic screening. There are lots of different arguments — so that means there are no straightforward answers.

More Genetic Diagrams

Predicting and explaining the outcomes of crosses between individuals is much easier when you've got a genetic diagram. So here are a couple more examples for you.

All the Offspring are Normal

Let's take another look at the <u>crazy hamster</u> example from page 57:

In this cross, a hamster with <u>two dominant alleles</u> (BB) is crossed with a hamster with <u>two recessive alleles</u> (bb). <u>All</u> the offspring are normal (boring).

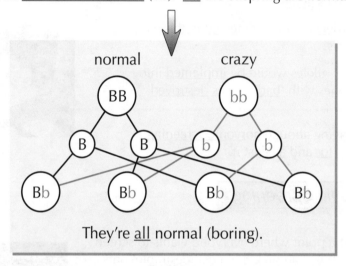

They're <u>all</u> normal (boring).

But, if you crossed a hamster with <u>two dominant alleles</u> (BB) with a hamster with <u>a dominant and a recessive allele</u> (Bb), you would also get <u>all</u> normal (boring) offspring.

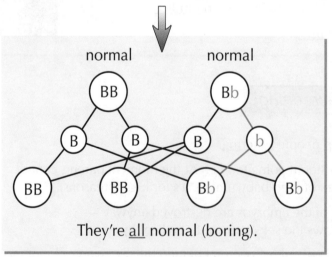

They're <u>all</u> normal (boring).

To find out <u>which</u> it was you'd have to <u>breed the offspring together</u> and see what kind of <u>ratio</u> you got that time — then you'd have a good idea. If it was <u>3:1</u>, it's likely that you originally had BB and bb.

More Genetic Diagrams

One more example of a genetic cross diagram coming up on this page.
Then a little bit about another type of genetic diagram — called a <u>family tree</u>...

There's a *1:1 Ratio* in the Offspring

1) A cat with <u>long hair</u> was bred with another cat with <u>short hair</u>.

2) The long hair is caused by a <u>dominant</u> allele 'H', and the short hair by a <u>recessive</u> allele 'h'.

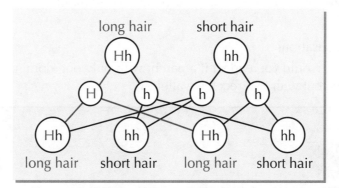

3) They had 8 kittens — 4 with long hair and 4 with short hair.

4) This is a <u>1:1</u> ratio — it's what you'd expect when a parent with only <u>one dominant allele</u> (Hh) is crossed with a parent with <u>two recessive alleles</u> (hh).

Family Trees Show How *Genetic Disorders* are *Inherited*

Knowing how inheritance works can help you to interpret a <u>family tree</u> — this is one for <u>cystic fibrosis</u>.

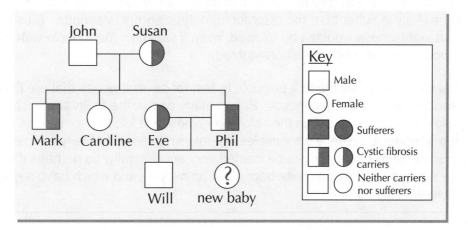

1) From the family tree, you can tell that the allele for cystic fibrosis <u>isn't</u> dominant because plenty of the family <u>carry</u> the allele but <u>aren't sufferers</u>.

2) There is a <u>25%</u> chance that the new baby will be a sufferer and a <u>50%</u> chance that it will be a carrier, as both of its parents are carriers but not sufferers. The case of the new baby is just the same as the genetic diagram on page 58 — so the baby could be <u>normal</u> (FF), a <u>carrier</u> (Ff) or a <u>sufferer</u> (ff).

It's enough to make you go cross-eyed...

If you see a <u>family tree</u> showing the inheritance of a <u>dominant allele</u>, it <u>won't</u> have any <u>carriers</u> on it (everyone who carries the allele is a sufferer). A good way to work out a family tree is to write the <u>genotype</u> of each person onto it — try copying the one above and writing the genotypes on for practice.

Warm-Up and Exam Questions

There's no better preparation for exam questions than doing... err... practice exam questions.
Hang on, what's this I see...

Warm-Up Questions

1) Which chromosomes decide whether you're male or female?
2) What are alleles?
3) Explain how it's possible to be a carrier of cystic fibrosis without knowing.
4) What is polydactyly?
5) What is *in vitro* fertilisation?
6) What offspring ratio would you expect if a parent with only one dominant allele
   is crossed with a parent with two recessive alleles?

Exam Questions

1   Read this passage about embryo screening.

> Embryo screening already happens in the UK for genetic disorders like cystic fibrosis.
> A genetic test can be done during IVF treatment — so that doctors can select a healthy
> embryo to implant in the mother. The other embryos are discarded.
>
> At the moment, regulations say that embryo screening is only allowed when there is
> "a significant risk of a serious genetic condition being present in the embryo."
> In other words, it is only allowed when a child of the person carrying the faulty allele
> would be likely to suffer from the disorder, and the disorder is serious. So screening
> for short-sightedness wouldn't be allowed, even if we knew that people with a faulty
> allele would definitely become short-sighted.
>
> Medical technology has made it possible to test for several genes that are linked to
> very serious illnesses, e.g. cancers. But, in many cases, the faulty allele isn't certain
> to cause cancer — it increases the risk, sometimes by a lot, but sometimes just slightly.
> So, at the moment, screening for alleles like this isn't allowed. Many people say this is
> right. Some kinds of cancer can be treated very successfully, so perhaps it's wrong to
> destroy embryos that might never become ill anyway — and which have a good chance
> of recovery if they do.

(a) Could embryos be screened for polydactyly under the current regulations?
    Explain your answer.

(1 mark)

(b) Cancer is a serious illness that kills thousands of people each year in the UK.
    Why is cancer not included in embryo screening?

(3 marks)

(c) Give **one** reason why a person might be opposed to
    screening embryos for any genetic condition.

(1 mark)

Exam Questions

2   In one of Gregor Mendel's experiments, he crossed thoroughbred purple-flowered pea plants with thoroughbred white-flowered plants. The first generation of offspring were all purple-flowered.

    (a) In Mendel's experiment, which characteristic is recessive?

(1 mark)

    (b) Using the symbols **F** and **f** to represent the alleles for purple and white, write down the combination of alleles (genotype) of each of the following:

       (i)   the original purple-flowered parent plant

(1 mark)

       (ii) the original white-flowered parent plant

(1 mark)

       (iii) the first generation of purple-flowered offspring

(1 mark)

3   Cystic fibrosis is a disease caused by a recessive allele.

          F = the normal allele
          f = the faulty allele that leads to cystic fibrosis

The genetic diagram below shows the possible inheritance of cystic fibrosis from one couple.

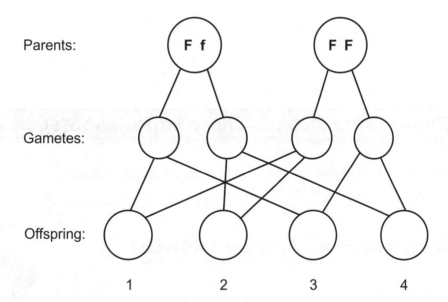

    (a) Copy and complete the genetic diagram.

(2 marks)

    (b) Which of the possible offspring will be sufferers and which will be unaffected?

(1 mark)

    (c) (i)   What proportion of the possible offspring are homozygous?

(1 mark)

       (ii) Which of the possible offspring are carriers of the disease?

(1 mark)

Genetic Engineering

Genetic engineering — playing around with genes. Cool.

Genetic Engineering *is Great* — *Hopefully*

The basic idea behind genetic engineering is to move genes (sections of DNA) from one organism to another so that it produces useful biological products. There are advantages and risks involved in genetic engineering.

Advantages

The main advantage is that you can produce organisms with new and useful features very quickly. There are some examples of this on the next page.

Risks

The main risk is that the inserted gene might have unexpected harmful effects. For example, genes are often inserted into bacteria so they produce useful products. If these bacteria mutated and became pathogenic (disease-causing), the foreign genes might make them more harmful and unpredictable. People also worry about the engineered DNA 'escaping' — e.g. weeds could gain rogue genes from a crop that's had genes for herbicide resistance inserted into it. Then they'd be unstoppable.

Genetic Engineering *Involves These* Important Stages:

1) First the gene that's responsible for producing the desirable characteristic is selected (say the gene for human insulin).

2) It's then 'cut' from the DNA using enzymes, and isolated.

3) The useful gene is inserted into the DNA of another organism (e.g. a bacterium).

4) The organism then replicates and soon there are loads of similar organisms all producing the same thing (e.g. loads of bacteria producing human insulin).

Genetic Engineering

On the face of it, genetic engineering is great.
But like lots of things, there are plenty of issues to consider.

Genetic Engineering has Many Uses

Here are three examples of how genetic engineering is used:

1) In some parts of the world, the population relies heavily on <u>rice</u> for food. In these areas, <u>vitamin A deficiency</u> can be a problem, because rice doesn't contain much of this vitamin, and other sources are <u>scarce</u>. Genetic engineering has allowed scientists to take a <u>gene</u> that controls <u>beta-carotene production</u> from <u>carrot plants</u>, and put it into <u>rice plants</u>. Humans can then change the beta-carotene into Vitamin A.

2) The gene for <u>human insulin production</u> has been put into <u>bacteria</u>. These are <u>cultured</u> in a <u>fermenter</u>, and the human insulin is simply <u>extracted</u> from the medium as they produce it.

3) Some plants have <u>resistance</u> to things like <u>herbicides</u>, <u>frost damage</u> and <u>disease</u>. Unfortunately, it's not always the plants we <u>want to grow</u> that have these features. But now, thanks to genetic engineering, we can <u>cut out</u> the gene responsible and stick it into <u>useful plants</u> such as <u>crops</u>.

There are Moral and Ethical Issues Involved

All this is nice, but there are <u>moral and ethical issues</u> surrounding genetic modification:

1) Some people think it's <u>wrong</u> to genetically engineer other organisms purely for <u>human benefit</u>. This is a particular problem in the genetic engineering of <u>animals</u>, especially if the animal <u>suffers</u> as a result.

2) People worry that we won't <u>stop</u> at engineering <u>plants</u> and <u>animals</u>. In the future, those who can afford genetic engineering might be able to decide the characteristics they want their <u>children</u> to have — and those who can't afford it may become a 'genetic underclass'.

3) The <u>evolutionary consequences</u> of genetic engineering are <u>unknown</u>, so some people think it's <u>irresponsible</u> to carry on when we're not sure what the <u>impact</u> on <u>future generations</u> might be.

Genetic engineering has exciting and frightening possibilities

It's up to the <u>Government</u> to weigh up all the <u>evidence</u> before <u>making a decision</u> on how this knowledge is used. All scientists can do is make sure the Government has all the information it needs.

Cloning Mammals

If you've cloned a sheep before then you won't need to look at this page. If not, you'd better <u>read on</u>...

Cloned Mammals Can be Made by Adult Cell Cloning

<u>Cloning</u> is a type of <u>asexual reproduction</u> (see page 46). It produces cells that are <u>genetically identical</u> to an original cell. Here's how it's done:

1) <u>Adult cell cloning</u> involves taking an unfertilised <u>egg cell</u> and removing its <u>haploid nucleus</u> (the egg cell is <u>enucleated</u>).

2) A nucleus is taken from an <u>adult body cell</u> (e.g. skin cell). This is a <u>diploid nucleus</u> containing the full number of <u>chromosomes</u>.

3) The diploid nucleus is <u>inserted</u> into the '<u>empty</u>' egg cell.

4) The egg cell is then stimulated by an <u>electric shock</u> — this makes it <u>divide</u> by mitosis, like a normal embryo.

5) When the embryo is a ball of cells, it's <u>implanted</u> into an <u>adult female</u> (the surrogate mother) to grow into a genetically identical copy (clone) of the original adult body cell.

6) This technique was used to create <u>Dolly</u> — the famous <u>cloned sheep</u>.

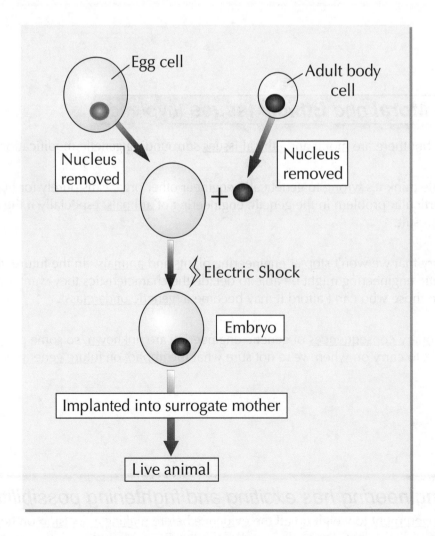

Cloning Mammals

Cloning isn't just used for sheep — there are many uses for it.

Cloning Has Many Uses

Cloning has several uses — here are just two examples:

1) Cloning mammals could help with the shortage of organs for transplants. For example, genetically-modified pigs are being bred that could provide suitable organs for humans. If this is successful, then cloning these pigs could help to meet the demand for organ transplants.

2) Human embryos could be produced by cloning adult body cells. The embryos could then be used to supply stem cells for stem cell therapy (see page 50). These cells would have exactly the same genetic information as the patient, reducing the risk of rejection (a common problem with transplants).

There are Many Issues Surrounding Cloning

1) Cloning mammals leads to a "reduced gene pool" — this means there are fewer different alleles in a population.

   - If a population are all closely related and a new disease appears, they could all be wiped out — because there may be no allele in the population giving resistance to the disease.

2) Cloned mammals mightn't live as long — Dolly the sheep only lived for 6 years (half as long as many sheep).

   - She was put down because she had lung disease, and she also had arthritis. These diseases are more usual in older sheep.
   - Dolly was cloned from an older sheep, so it's been suggested her 'true' age may have been older.
   - But it's possible she was just unlucky — and that her illnesses weren't linked to her being a clone.

3) There are other risks and problems associated with cloning:

   - The cloning process often fails. It took hundreds of attempts to clone Dolly.
   - Clones are often born with genetic defects.
   - Cloned mammals' immune systems are sometimes unhealthy — so they suffer from more diseases.

Cloning produces genetically identical organisms

Cloning can be a controversial topic — especially when it's to do with cloning animals (and especially humans). Is it healthy scientific progress, or are we trying to 'play God'?

Selective Breeding

'Selective breeding' sounds like it has the potential to be a tricky topic, but it's actually dead simple. You take the <u>best</u> plants or animals and breed them together to get the best possible <u>offspring</u>. That's it.

Selective Breeding is Very Simple

Selective breeding is when humans artificially select the plants or animals that are going to <u>breed</u> and have their genes remain in the population, according to what <u>we</u> want from them. Organisms are <u>selectively bred</u> to develop the <u>best features</u>, which are things like:

- <u>Maximum yield</u> of meat, milk, grain etc.
- <u>Good health</u> and <u>disease resistance</u>.
- Other qualities like <u>temperament</u>, <u>speed</u>, <u>attractiveness</u>, etc.

This is the basic process involved in <u>selective breeding</u>:

1) From your <u>existing stock</u> select the ones which have the <u>best characteristics</u>.
2) <u>Breed them</u> with each other.
3) Select the <u>best</u> of the <u>offspring</u>, and <u>breed them together</u>.
4) Continue this process over <u>several generations</u>, and the desirable trait gets <u>stronger</u> and <u>stronger</u>.

Example

In <u>agriculture</u> (farming), selective breeding can be used to <u>improve yields</u>. E.g. to improve <u>meat yields</u>, a farmer could breed together the <u>cows</u> and <u>bulls</u> with the <u>best characteristics</u> for producing <u>meat</u>, e.g. large size. After doing this for <u>several generations</u> the farmer would get cows with a <u>very high meat yield</u>.

The **Main Drawback** is a **Reduction** in the **Gene Pool**

1) The main problem with selective breeding is that it reduces the <u>gene pool</u> in a population (see previous page). This is because the farmer keeps breeding from the "<u>best</u>" animals or plants — which are all <u>closely related</u>. This is known as <u>inbreeding</u>.

2) Inbreeding can cause <u>health problems</u> because there's more chance of the organisms developing <u>harmful genetic disorders</u> when the <u>gene pool</u> is <u>limited</u>. This is because lots of genetic conditions are <u>recessive</u> — you need <u>two alleles</u> to be the <u>same</u> for it to have an effect. Breeding from closely related organisms all the time means that <u>recessive alleles</u> are <u>more likely</u> to <u>build up</u> in the population (because the organisms are likely to share the same alleles).

3) There can also be serious problems if a <u>new disease appears</u>, because there's <u>not much variation</u> in the population (see page 48). All the stock are <u>closely related</u> to each other, so if one of them is going to be killed by a new disease, the others are <u>also</u> likely to succumb to it.

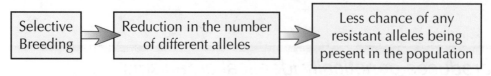

Warm-Up and Exam Questions

By doing these warm-up questions, you'll soon find out what you know and what you don't. Once you've finished, take the time to go back over the bits you've struggled with.

Warm-Up Questions

1) Briefly explain the basic idea behind genetic engineering.
2) Name one useful product that humans have genetically modified bacteria to produce.
3) What is a clone?
4) In adult cell cloning, how is the egg cell stimulated to divide?
5) What is a reduced gene pool?
6) How is selective breeding done?

Exam Questions

1 Organisms can be genetically modified.
This means an organism's genes can be altered to alter its characteristics.

(a) Give **one** function of the enzymes that are used in genetic engineering.

(1 mark)

(b) Suggest **one** useful way that plants can be genetically modified.

(1 mark)

(c) Give three reasons why some people are opposed to genetic modification.

(3 marks)

2 In 1997 scientists at the Roslin Institute announced the birth of Dolly the sheep, the first mammal to be cloned using adult cell cloning.

(a) Explain how a sheep can be cloned using adult cell cloning.

(4 marks)

(b) Dolly was a female sheep. What would have been the gender of the sheep that Dolly was created from? Explain your answer.

(1 mark)

Exam Questions

3   The diagram shows a method that scientists used for cloning mice.
     They used three adult mice for this — Beatrix, Brenda and Belinda.

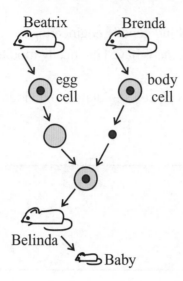

(a) Explain why the egg cell from Beatrix had its nucleus removed.

(1 mark)

(b) Explain why the nucleus was taken from Brenda's body cell.

(1 mark)

(c) One of the three adult mice had a useful property which the scientists
     wanted to clone. Which of them do you think this was? Explain your answer.

(1 mark)

4   The Chestnut Stud Farm breeds racehorses. Over the years, each generation of horses was
     faster than their parents.
     (a) Explain how this could be achieved as a result of a breeding programme.

(3 marks)

     (b) Suggest **one** disadvantage of such a breeding programme.

(1 mark)

5   Cloning is a controversial issue which many people feel passionate about.
     (a) Describe **two** uses of cloning.

(2 marks)

*In this question you will be assessed on the quality of your English, the organisation
of your ideas and your use of appropriate specialist vocabulary.*

     (b) Discuss the potential disadvantages of cloning technology.

(6 marks)

Fossils and Extinction

Fossils can be really useful — if they're <u>well preserved</u>, they can show you what creatures that have been dead for millions of years might have <u>looked</u> like.

Fossils are the **Remains** *of Plants and Animals*

Fossils are the <u>remains</u> of organisms from <u>many years ago</u>, which are found in <u>rocks</u>.
Fossils provide the <u>evidence</u> that organisms lived ages ago.

Fossils form in rocks in one of <u>three</u> ways:

1) From **Gradual Replacement** *by* **Minerals**

Most fossils happen this way...

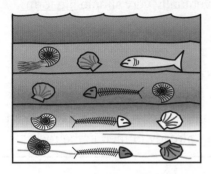

1) Things like <u>teeth</u>, <u>shells</u>, <u>bones</u> etc., which <u>don't decay</u> easily, can last a long time when <u>buried</u>.

2) They're eventually <u>replaced by minerals</u> as they decay, forming a <u>rock-like substance</u> shaped like the original hard part.

3) The surrounding sediments also turn to rock, but the fossil stays <u>distinct</u> inside the rock and eventually someone <u>digs it up</u>.

2) From **Casts** *and* **Impressions**

1) Sometimes, fossils are formed when an organism is <u>buried</u> in a <u>soft</u> material like clay. The clay later <u>hardens</u> around it and the organism decays, leaving a <u>cast</u> of itself.

> An animal's <u>burrow</u> or a plant's <u>roots</u> can be preserved as casts.

2) Things like footprints can be <u>pressed</u> into these materials when soft, leaving an <u>impression</u> when it hardens.

3) From **Preservation** *in Places Where* **No Decay** *Happens*

1) In <u>amber</u> (a clear yellow 'stone' made from fossilised resin) and <u>tar pits</u> there's no <u>oxygen</u> or <u>moisture</u> so <u>decay microbes</u> can't survive.

2) In <u>glaciers</u> it's too <u>cold</u> for the <u>decay microbes</u> to work.

3) <u>Peat bogs</u> are too <u>acidic</u> for <u>decay microbes</u>.

Fossils give us information about ancient animals and plants

It's amazing that <u>fossils</u> of organisms can still exist millions of years after they died. There are <u>three ways</u> that fossils form and all the details are right here on this page — aren't you lucky...

Fossils and Extinction

Fossils can also help us to work out how life on Earth has <u>evolved</u>. Evolution leads to the development of lots of <u>different species</u>. But not every species is still around today...

No One Knows How Life Began...

Fossils show how many of today's species have <u>evolved</u> (changed and developed) over millions of years. But where did the <u>first</u> living thing come from...

1) There are various <u>hypotheses</u> suggesting how life first came into being, but no one really <u>knows</u>.

2) Maybe the first life forms came into existence in a primordial <u>swamp</u> (or under the <u>sea</u>) here on <u>Earth</u>. Maybe simple organic molecules were brought to Earth on <u>comets</u> — these could have then become more <u>complex</u> organic molecules, and eventually very simple <u>life forms</u>.

3) These hypotheses can't be supported or disproved because there's a <u>lack</u> of <u>valid</u> and <u>reliable</u> evidence.

   Validity and reliability are explained on page 3.

4) There's a lack of evidence because scientists believe many early organisms were <u>soft-bodied</u>, and soft tissue tends to decay away <u>completely</u>. So the fossil record is <u>incomplete</u>.

5) Plus, fossils that did form millions of years ago may have been <u>destroyed</u> by <u>geological activity</u>, e.g. the movement of tectonic plates may have crushed fossils already formed in the rock.

Extinction Happens if You Can't Evolve Quickly Enough

The fossil record contains many species that <u>don't exist any more</u> — these species are said to be <u>extinct</u>. <u>Dinosaurs</u> and <u>mammoths</u> are extinct animals, with only <u>fossils</u> to tell us they existed at all.

Species become extinct for these reasons:

1) The <u>environment changes</u> too quickly (e.g. destruction of habitat).

2) A <u>new predator</u> kills them all (e.g. humans hunting them).

3) A <u>new disease</u> kills them all.

4) They can't <u>compete</u> with another (new) species for <u>food</u>.

5) A <u>catastrophic event</u> happens that kills them all (e.g. a volcanic eruption or a collision with an asteroid).

6) A <u>new species</u> develops (this is called speciation — see next page).

Dodos are now extinct. Humans not only hunted them, but introduced other animals which ate all their eggs, and we destroyed the forest where they lived — they really didn't stand a chance...

Species evolve — or become extinct...

We don't really know how life on Earth began — there are plenty of hypotheses, but we won't really know unless we find valid and reliable evidence. I'm afraid that's just science for you...

Speciation

If you've been wondering how a <u>new species</u> can spring up, this is the page for you.

Speciation is the Development of a New Species

1) A species is a group of <u>similar organisms</u> that can <u>reproduce</u> to give <u>fertile offspring</u>.

2) <u>Speciation</u> is the development of a <u>new species</u>.

3) Speciation occurs when <u>populations</u> of the <u>same species</u> become so <u>different</u> that they can <u>no longer breed</u> together to produce <u>fertile offspring</u>.

Isolation and Natural Selection Lead to Speciation

<u>ISOLATION</u> is where <u>populations</u> of a species are <u>separated</u>.

1) Isolation can happen due to a <u>physical barrier</u>. E.g. floods and earthquakes can cause barriers that <u>geographically isolate</u> some individuals from the main population.

2) <u>Conditions</u> on either side of the barrier will be <u>slightly different</u>, e.g. they may have <u>different climates</u>.

3) Because the environment is <u>different</u> on each side, <u>different characteristics</u> will become more common in each population due to <u>natural selection</u>:

- Each population shows <u>variation</u> because they have a wide range of <u>alleles</u>.
- In each population, individuals with characteristics that make them better adapted to their environment have a <u>better chance of survival</u> and so are more likely to <u>breed</u> successfully.
- So the <u>alleles</u> that control the <u>beneficial characteristics</u> are more likely to be <u>passed on</u> to the <u>next generation</u>.

4) Eventually, individuals from the different populations will have <u>changed</u> so much that they <u>won't</u> be able to <u>breed</u> with one another to produce fertile offspring.

5) The two groups will have become <u>separate species</u>:

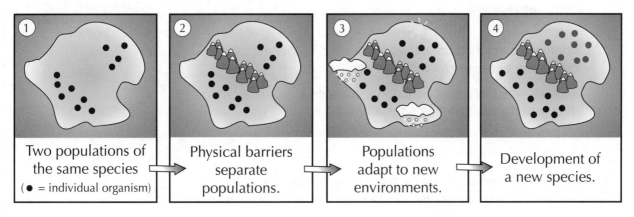

| Two populations of the same species (● = individual organism) | Physical barriers separate populations. | Populations adapt to new environments. | Development of a new species. |

A new species can develop when populations become separated

So <u>speciation</u> happens if two or more populations of the same species change so much that they can <u>no longer breed together</u> to produce <u>fertile offspring</u>. It can be caused by populations becoming <u>separated</u> from each other. Get your head around this stuff, then it's time for some more questions...

Warm-Up and Exam Questions

You need to test your knowledge with a few warm-up questions, followed by some exam questions...

Warm-Up Questions

1) What are fossils?
2) Give an example of something that can be preserved as a cast.
3) Some fossils of early organisms did form in ancient rocks.
   Explain why many of these fossils have not survived to the present day.
4) What is an extinct species?
5) What does the term 'isolation' mean in the context of speciation?

Exam Questions

1   The fossil of an extinct species of insect is found preserved inside a piece of amber.

   (a)   Explain why the remains of the insect have been preserved
         so well inside the amber.

                                                                                    (2 marks)

   (b)   This particular species of insect became extinct because a catastrophic
         event wiped out every member of the species at once.

         Give **one** example of the kind of catastrophic
         event that could wipe out a species.

                                                                                    (1 mark)

2   The picture on the right shows what scientists
     believe the dinosaur Stegosaurus looked like.

   (a)   Fossils of Stegosaurus teeth have been discovered.
         Briefly explain how the fossils of the Stegosaurus
         teeth were formed.

                                                                                    (2 marks)

   (b)   Scientists cannot be completely sure what Stegosaurus
         looked like because of a lack of evidence.

         Suggest why there is not enough evidence to
         show exactly what Stegosaurus looked like.

                                                                                    (1 mark)

3   Two different species of birds are found on two nearby islands.
     The two species were originally just one species living on one of the islands.

     Suggest an explanation for how the two different bird species developed.

                                                                                    (5 marks)

Revision Summary for Section 2

Wow, that was quite a long section. First there was all the stuff about enzymes and then came all the genetics bits. And just to finish off, some questions. Use these to find out what you know about it all — and what you don't. Then look back and learn the bits you don't know.
Then try the questions again, and again...

1) What shape is a molecule of DNA?
2) Name the four scientists who had major roles in discovering the structure of DNA.
3) Explain how DNA fingerprinting is used in forensic science.
4) What does a triplet of DNA bases code for?
5) Are mutations always harmful? Explain your answer.
6) Describe the stages of protein synthesis.
7) What is mitosis used for in the human body? Describe the four steps in mitosis.
8) Where does meiosis take place in the human body?
9) What is differentiation in a cell?
10) Give three ways that embryonic stem cells could be used to cure diseases.
11) Humans go through two main phases of rapid growth. When do these take place?
12) Copy and complete the diagrams to show what happens to the X and Y chromosomes during reproduction.

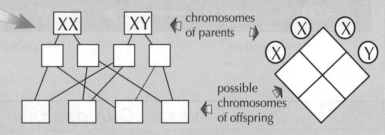

13) List three important conclusions that Mendel reached following his experiments with pea plants.
14) The significance of Mendel's work was not realised until 1900, 16 years after Mendel died. Suggest why the importance of the work wasn't understood at the time.
15) What is meant by an organism being heterozygous? What about homozygous?
16) Describe the basic difference between a recessive allele and a dominant one.
17) If both parents carry recessive allele for cystic fibrosis, what is the probability of their child being a carrier?
18)*Blue colour in a plant is carried on a recessive allele, b. The dominant allele, B, gives white flowers. In the first generation after a cross, all the flowers are white. These are bred together and the result is a ratio of 54 white : 19 blue. What were the alleles of the flowers used in the first cross?
19) Give one advantage of genetic engineering.
20) Describe three examples of genetic engineering.
21) What is selective breeding?
22) Describe the three ways that fossils can form. Give an example of each type.
23) Give three reasons why some species become extinct.
24) What is speciation? Explain how geographical isolation can lead to speciation.

* Answer on page 271.

Plant Structure and Photosynthesis

Plants carry out photosynthesis to produce food. You're about to find out all about it.

Plant Cells Are Organised Into Tissues And Organs

Plants are made of organs like stems, roots and leaves. These organs are made of tissues.
For example, leaves are made of:

- Mesophyll tissue — this is where most of the photosynthesis in a plant occurs.
- Xylem and phloem — they transport things like water, mineral ions and sucrose around the plant.
- Epidermal tissue — this covers the whole plant.

The leaf diagram at the bottom of the page shows where these tissues are in a plant.

Photosynthesis Has An Equation:

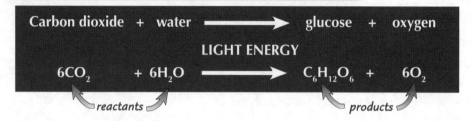

Carbon dioxide + water ⟶ glucose + oxygen

LIGHT ENERGY

$$6CO_2 + 6H_2O \longrightarrow C_6H_{12}O_6 + 6O_2$$

reactants   *products*

Photosynthesis Produces Glucose Using Sunlight

1) Photosynthesis is the process that produces 'food' in plants and algae.
   The 'food' it produces is glucose.

2) Photosynthesis happens inside the chloroplasts.

3) Chloroplasts contain a green substance called chlorophyll, which absorbs sunlight and uses
   its energy to convert carbon dioxide (from the air) and water (from the soil) into glucose.
   Oxygen is also produced as a by-product.

4) Photosynthesis happens in the leaves of all green plants — this is largely what the leaves are for.
   Below is a cross-section of a leaf showing the four raw materials needed for photosynthesis.

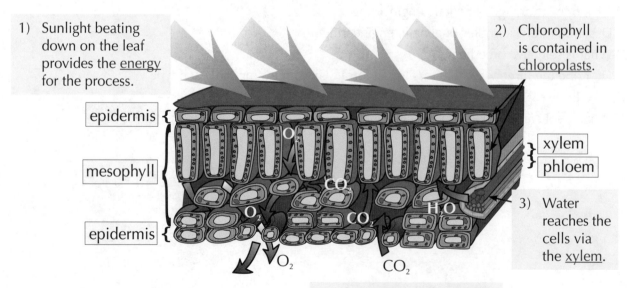

1) Sunlight beating
   down on the leaf
   provides the energy
   for the process.

2) Chlorophyll
   is contained in
   chloroplasts.

epidermis

mesophyll

epidermis

xylem

phloem

3) Water
   reaches the
   cells via
   the xylem.

4) CO₂ diffuses into the leaf.

The Rate of Photosynthesis

The rate of photosynthesis is affected by the intensity of <u>light</u>, the amount of <u>CO_2</u>, and the <u>temperature</u>. Plants also need <u>water</u> for photosynthesis, but when a plant is so short of water that it becomes the <u>limiting factor</u> in photosynthesis, it's already in such <u>trouble</u> that this is the least of its worries.

The **Limiting Factor** Depends on the Conditions

1) Any of the three factors above can become the <u>limiting factor</u>. This just means that it's stopping photosynthesis from happening any <u>faster</u>.

2) Which factor is limiting at a particular time depends on the <u>environmental conditions</u>:

   • at <u>night</u> it's pretty obvious that <u>light</u> is the limiting factor,

   • in <u>winter</u> it's often the <u>temperature</u>,

   • if it's warm enough and bright enough, the amount of <u>CO_2</u> is usually limiting.

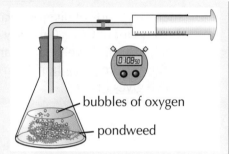

You can do <u>experiments</u> to work out the <u>ideal conditions</u> for photosynthesis in a particular plant. The easiest type to use is a water plant like <u>Canadian pondweed</u> — you can easily measure the amount of <u>oxygen produced</u> in a given time to show how <u>fast</u> photosynthesis is happening (remember, oxygen is made during photosynthesis).

You could either count the <u>bubbles</u> given off, or if you want to be a bit more <u>accurate</u> you could <u>collect</u> the oxygen in a <u>gas syringe</u>.

bubbles of oxygen

pondweed

Not Enough **Light** Slows Down the Rate of Photosynthesis

1) Light provides the <u>energy</u> needed for photosynthesis.

2) As the <u>light level</u> is raised, the rate of photosynthesis <u>increases steadily</u> — but only up to a <u>certain point</u>.

3) Beyond that, it <u>won't</u> make any difference because then it'll be either the <u>temperature</u> or the <u>CO_2 level</u> which is the limiting factor.

4) In the lab you can change the light intensity by <u>moving a lamp</u> closer to or further away from your plant.

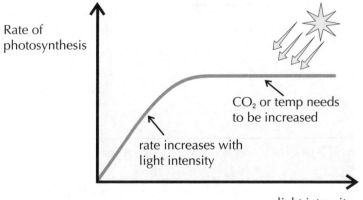

Rate of photosynthesis

CO_2 or temp needs to be increased

rate increases with light intensity

light intensity

5) But if you just plot the rate of photosynthesis against "distance of lamp from the beaker", you get a <u>weird-shaped graph</u>. To get a graph like the one above you either need to <u>measure</u> the light intensity at the beaker using a <u>light meter</u> or do a bit of nifty maths with your results.

The Rate of Photosynthesis

Too Little **Carbon Dioxide** Also Slows it Down

1) CO_2 is one of the <u>raw materials</u> needed for photosynthesis.

2) As with light intensity the amount of <u>CO_2</u> will only increase the rate of photosynthesis up to a point. After this the graph <u>flattens out</u> showing that CO_2 is no longer the <u>limiting factor</u>.

3) As long as <u>light</u> and <u>CO_2</u> are in plentiful supply then the factor limiting photosynthesis must be <u>temperature</u>.

4) There are loads of different ways to control the amount of CO_2. One way (for a water plant) is to dissolve different amounts of <u>sodium hydrogencarbonate</u> in the water, which <u>gives off</u> CO_2.

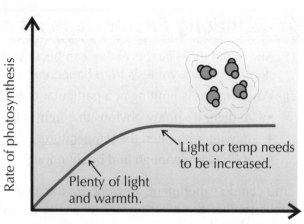

Light or temp needs to be increased.

Plenty of light and warmth.

% level of CO_2

The **Temperature** has to be Just Right

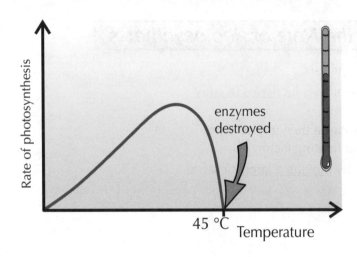

enzymes destroyed

45 °C Temperature

1) Usually, if the temperature is the <u>limiting factor</u> it's because it's <u>too low</u> — the <u>enzymes</u> needed for photosynthesis work more <u>slowly</u> at low temperatures.

2) But if the plant gets <u>too hot</u>, the enzymes it needs for photosynthesis and its other reactions will be <u>damaged</u>.

3) This happens at about <u>45 °C</u> (which is pretty hot for outdoors, although <u>greenhouses</u> can get that hot if you're not careful).

4) Experimentally, the best way to control the temperature of the flask is to put it in a <u>water bath</u>.

In all these experiments, you have to try and keep all the variables <u>constant</u> (apart from the one you're investigating) so it's a <u>fair test</u>:

- use a <u>bench lamp</u> to control the intensity of the light (careful not to <u>block the light</u> with anything)

- keep the flask in a <u>water bath</u> to help keep the temperature constant

- you <u>can't</u> really do anything about the CO_2 levels — you just have to use a <u>large flask</u>, and do the experiments as <u>quickly</u> as you can, so that the plant doesn't use up too much of the CO_2 in the flask. If you're using sodium hydrogencarbonate make sure it's changed each time.

The Rate of Photosynthesis

Growing plants outdoors can be <u>very difficult</u>, especially on a <u>large scale</u> — it's almost impossible to control the weather and other conditions. But there's a way around that...

You can *Artificially Create* the *Ideal Conditions* for *Farming*

1) The most common way to artificially create the <u>ideal environment</u> for plants is to grow them in a <u>greenhouse</u>.

2) Greenhouses help to <u>trap</u> the Sun's <u>heat</u>, and make sure that the <u>temperature</u> doesn't become <u>limiting</u>. In winter a farmer or gardener might use a <u>heater</u> as well to keep the temperature at the ideal level. In summer it could get <u>too hot</u>, so they might use <u>shades</u> and <u>ventilation</u> to cool things down.

3) <u>Light</u> is always needed for photosynthesis, so commercial farmers often supply <u>artificial light</u> after the Sun goes down to give their plants more quality photosynthesis time.

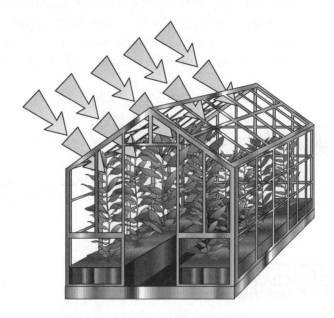

Greenhouses are used to grow plants, including food crops, flowers and tobacco plants.

4) Farmers and gardeners can also increase the level of <u>carbon dioxide</u> in the greenhouse. A fairly common way is to use a <u>paraffin heater</u> to heat the greenhouse. As the paraffin burns, it makes carbon dioxide as a <u>by-product</u>.

5) Keeping plants <u>enclosed</u> in a greenhouse also makes it easier to keep them free from <u>pests</u> and <u>diseases</u>. The farmer can add <u>fertilisers</u> to the soil as well, to provide all the <u>minerals</u> needed for healthy growth.

6) Sorting all this out <u>costs money</u> — but if the farmer can keep the conditions <u>just right</u> for photosynthesis, the plants will grow much <u>faster</u> and a <u>decent crop</u> can be harvested much more <u>often</u>, which can then be <u>sold</u>. It's important that a farmer supplies just the <u>right amount</u> of heat, light, etc. — enough to make the plants grow well, but <u>not</u> more than the plants <u>need</u>, as this would just be <u>wasting money</u>.

Greenhouses control the growing environment

Farmers use greenhouses to make sure crops get the right amount of carbon dioxide, light and heat. They can alter the conditions using paraffin heaters, artificial light and ventilation. This ensures nothing becomes a limiting factor for photosynthesis, which means a good crop is produced.

How Plants Use Glucose

Once plants have made the glucose, there are various ways they can use it.

1) For **Respiration**

1) Plants manufacture glucose in their leaves.
2) They then use some of the glucose for respiration (see page 28).
3) This releases energy which enables them to convert the rest of the glucose into various other useful substances, which they can use to build new cells and grow.
4) To produce some of these substances they also need to gather a few minerals from the soil.

2) Making **Cell Walls**

Glucose is converted into cellulose for making strong cell walls (see page 13), especially in a rapidly growing plant.

Algae also use glucose to make cellulose for cell walls, fats and oils for storage, and amino acids for proteins.

3) Making **Proteins**

Glucose is combined with nitrate ions (absorbed from the soil) to make amino acids, which are then made into proteins.

4) Stored in **Seeds**

Glucose is turned into lipids (fats and oils) for storing in seeds. Sunflower seeds, for example, contain a lot of oil — we get cooking oil and margarine from them. Seeds also store starch (see below).

5) Stored as **Starch**

1) Glucose is turned into starch and stored in roots, stems and leaves, ready for use when photosynthesis isn't happening, like in the winter.
2) Starch is insoluble which makes it much better for storing than glucose — a cell with lots of glucose in would draw in loads of water and swell up.
3) Potato and parsnip plants store a lot of starch underground over the winter so a new plant can grow from it the following spring. We eat the swollen storage organs.

All life depends on photosynthesis

Plants are pretty crucial in ensuring the flow of energy through nature. They are able to use the Sun's energy to make glucose, the energy source that humans and other animals need for respiration. Make sure you know the photosynthesis equation inside out — look back at p. 76 if you don't.

Warm-Up and Exam Questions

So, here we go again — another set of questions to test your knowledge. But don't roll your eyes, I promise they'll be really, really enjoyable. OK, don't hold me to that, but make sure you do them...

Warm-Up Questions

1) Name the four raw materials that are needed for photosynthesis.
2) What is meant by a limiting factor for the rate of photosynthesis?
3) Sketch a graph to show how the rate of photosynthesis varies with increasing CO_2.
4) Give one way in which a farmer could increase the level of CO_2 in his greenhouse.
5) Name the process that converts glucose into energy in plants.

Exam Questions

1 Photosynthesis makes glucose. The glucose may then be converted to other substances. Some of these substances are listed below:

   starch   **cellulose**   **amino acids**

   Match each of these substances to its correct function from the list below.
   • Making cell walls.
   • Making proteins.
   • Storing energy.

   (3 marks)

2 The diagram shows part of the structure of a leaf as it looks under a microscope.

   (a) Name the tissue labelled **A**.

   (1 mark)

   (b) Name **two** other tissues that can be found in a leaf.

   (2 marks)

3 The table shows the rate of photosynthesis of a plant at different temperatures.

Temperature (°C)	Rate of photosynthesis (arbitrary units)
0	0
10	17
20	35
30	67
40	82
50	0

   (a) Explain the difference between:

   (i) the rates of photosynthesis at 10 °C and at 20 °C.

   (1 mark)

   (ii) the rates of photosynthesis at 40 °C and at 50 °C.

   (1 mark)

   (b) A student said that the optimum temperature for photosynthesis in this plant was 40 °C. Comment on this statement.

   (2 marks)

Exam Questions

4   Jane did an experiment to see how the rate of photosynthesis depends on light intensity.
The diagram shows her apparatus.

(a)   How can Jane measure the rate of photosynthesis?

(1 mark)

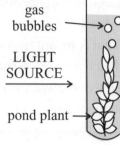

(b)   In this experiment:

(i)   what is the dependent variable?

(1 mark)

(ii)   what is the independent variable?

(1 mark)

(c)   State **one** factor that should be kept constant
during this experiment.

(1 mark)

5   The graph shows how a plant's rate of
photosynthesis varies with the light intensity.

(a)   Label a point at which light intensity is
the limiting factor for photosynthesis.

(1 mark)

(b)   What else can limit a plant's
rate of photosynthesis?

(1 mark)

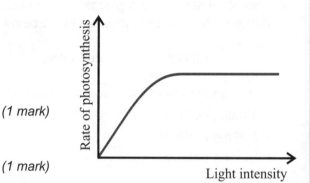

6   The diagram shows a variegated leaf —
it's partly green and partly white.

(a)   What substance is present in the green
parts of the leaf but not the white parts?

(1 mark)

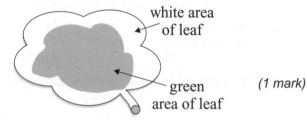

Abisola did an experiment in which part of the
leaf was covered with black paper, as shown:
The leaf was then exposed to light for four hours and was then tested for starch.

(b)   (i)   Copy and complete the diagram below by shading in the part(s) of the
leaf that you would expect to contain **starch**.

(1 mark)

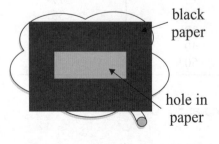

(ii)   Explain your answer to part (b)(i).

(3 marks)

Root Hairs and Active Transport

If you don't water a house plant for a few days it starts to go all droopy. Plants need <u>water</u>.

Root Hairs Take in Water by Osmosis

1) The cells on plant roots grow into long '<u>hairs</u>' which stick out into the soil.
   Each branch of a root will be covered in <u>millions</u> of these microscopic hairs.

2) This gives the plant a <u>big surface area</u> for <u>absorbing water</u> from the soil.

3) There's usually a <u>higher concentration</u> of water in the soil than there is inside the plant,
   so the water enters the root hair cell by <u>osmosis</u> (see page 32).

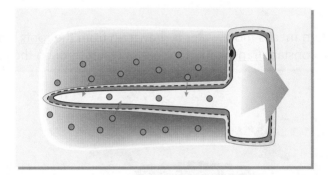

Root Hairs Take In Minerals Using Active Transport

1) <u>Root hairs</u> also absorb minerals from the soil.

2) But the <u>concentration</u> of minerals in the <u>soil</u> is usually pretty <u>low</u>.
   It's normally <u>higher</u> in the <u>root hair cell</u> than in the soil around it.

3) So normal diffusion <u>doesn't</u> explain how minerals are taken up into the root hair cell.

4) The answer is that a different process called '<u>active transport</u>' is responsible.

5) Active transport uses <u>energy</u> from <u>respiration</u> to help the plant pull minerals into the
   root hair <u>against the concentration gradient</u>. This is essential for its growth.

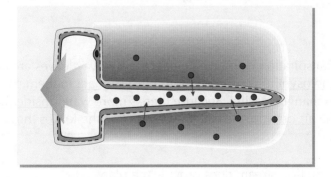

Plants have <u>tube networks</u> to move substances to and from individual cells quickly:

- <u>XYLEM</u> tubes transport <u>water and minerals</u> from the <u>root</u> to the <u>rest of the plant</u> (e.g. the <u>leaves</u>).
- <u>PHLOEM</u> tubes transport <u>sugars</u> from the <u>leaves</u> (where they're made) to <u>growing</u> and
  <u>storage</u> tissues.

Transpiration

You've just read about <u>water</u> entering a plant through the root hairs. Now for its <u>exit</u>...

Transpiration is the Loss of Water from the Plant

1) Transpiration is caused by the <u>evaporation</u> and <u>diffusion</u> (see page 31) of water from inside the leaves.

2) This creates a slight <u>shortage</u> of water in the leaf, and so more water is drawn up from the rest of the plant through the <u>xylem vessels</u> to replace it.

3) This in turn means more water is drawn up from the <u>roots</u>, and so there's a constant <u>transpiration stream</u> of water through the plant.

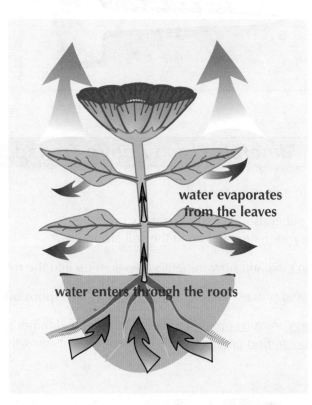

water evaporates from the leaves

water enters through the roots

4) Transpiration is just a <u>side-effect</u> of the way leaves are adapted for <u>photosynthesis</u>. They have to have <u>stomata</u> in them so that gases can be exchanged easily. Because there's more water <u>inside</u> the plant than in the <u>air outside</u>, the water escapes from the leaves through the stomata.

5) But, the transpiration stream does provide the plant with a constant supply of water for <u>photosynthesis</u>.

Transpiration involves evaporation and diffusion

A big tree loses about a <u>thousand litres</u> of water from its leaves <u>every single day</u> — it's a fact. That's as much water as the average person drinks in a whole year, so the <u>roots</u> have to be very effective at drawing in water from the soil. Which is why they have all those root <u>hairs</u>, you see.

Plant Development

Plants, like animals, have stem cells. Page 49 will refresh your memory on stem cells. Then read on...

Meristems Contain *Plant Stem Cells*

1) In plants, the only cells that are mitotically active (i.e. divide by mitosis) are found in plant tissues called meristems.

2) Meristem tissue is found in the areas of a plant that are growing — such as the roots and shoots.

3) Meristems produce unspecialised cells that are able to divide and form any cell type in the plant — they act like embryonic stem cells (see page 49). But unlike human stem cells, these cells can divide to generate any type of cell for as long as the plant lives.

4) The unspecialised cells can become specialised and form tissues like xylem and phloem (the water and food transport tissues).

5) These tissues can group together to form organs like leaves, roots, stems and flowers.

Clones of Plants Can be Produced from *Cuttings*

1) A cutting is part of a plant that has been cut off it.

2) Cuttings taken from an area of the plant that's growing will contain unspecialised meristem cells which can differentiate to make any cell.

3) This means a whole new plant can grow from the cutting which will be a clone of the parent plant.

4) Gardeners often take cuttings from parent plants with desirable characteristics, and then plant them to produce identical copies of the parent plant.

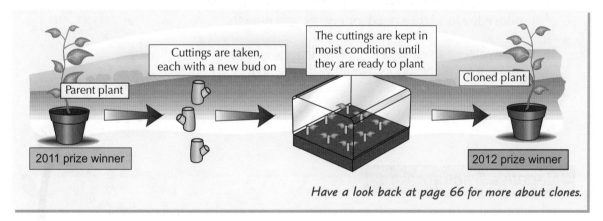

Parent plant

2011 prize winner

Cuttings are taken, each with a new bud on

The cuttings are kept in moist conditions until they are ready to plant

Cloned plant

2012 prize winner

Have a look back at page 66 for more about clones.

Rooting Powder Helps *Cuttings* to Grow into *Complete Plants*

1) If you stick cuttings in the soil they won't always grow.

2) If you add rooting powder, which contains plant hormones (auxins, see page 87) they'll produce roots rapidly and start growing as new plants.

3) This helps growers to produce lots of clones of a really good plant very quickly.

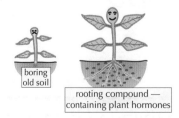

boring old soil

rooting compound — containing plant hormones

Phototropism

Unlike us humans, plants can't just get up and walk to something that they want. They can <u>grow</u> towards things though — <u>plant hormones</u> make sure they grow in a <u>useful direction</u> (e.g. toward light).

Phototropism is Growth Towards or Away From Light

1) Some parts of a plant, e.g. roots and shoots, can <u>respond</u> to <u>light</u> by <u>growing</u> in a certain <u>direction</u> — this is called <u>phototropism</u>.

2) Shoots are <u>positively phototropic</u> — they grow <u>towards</u> light.

3) Roots are <u>negatively phototropic</u> — they grow <u>away</u> from light.

Phototropism Helps Plants to Survive

Positive Phototropism

1) Plants need <u>sunlight</u> for <u>photosynthesis</u>.

2) Without sunlight, plants can't photosynthesise and don't produce the food they need for <u>energy and growth</u>.

3) Photosynthesis occurs <u>mainly</u> in the <u>leaves</u>, so it's important for plant shoots, which will grow leaves, to grow <u>towards light</u>.

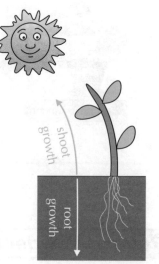

Negative Phototropism

1) Plants need <u>nutrients and water</u> from the <u>soil</u> to grow.

2) Phototropism means roots grow <u>away</u> from light, <u>down into the soil</u> where they can <u>absorb</u> the water and nutrients the plant needs for <u>healthy growth</u>.

Plants respond to their environments more than you might think

Plants need plenty of light to grow well. But they don't just sit around passively and hope for the best. No, thanks to auxins they can grow in the right directions to find what they need — see the next page.

Auxins

Auxins are responsible for helping plants grow in the <u>right direction</u> to get what they want.

Auxins are Plant Growth Hormones

1) <u>Auxins</u> are <u>chemicals</u> that control <u>growth</u> near the <u>tips</u> of <u>shoots</u> and <u>roots</u>.

2) Auxins are produced in the <u>tips</u> and <u>diffuse backwards</u> to stimulate the <u>cell elongation (enlargement) process</u>, which occurs in the cells <u>just behind</u> the tips.

3) If the tip of a shoot is <u>removed</u>, no auxins are available and the shoot may <u>stop growing</u>.

4) Auxins are involved in the responses of plants to <u>light</u>, <u>gravity</u> and <u>water</u>.

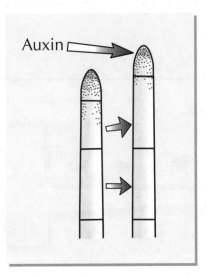

Auxins Make Shoots Grow Towards Light

1) When a <u>shoot tip</u> is exposed to <u>light</u>, <u>more auxins</u> accumulate on the side that's in the <u>shade</u> than the side that's in the light.

2) This makes the cells grow (elongate) <u>faster</u> on the <u>shaded side</u>, so the shoot grows <u>towards</u> the light.

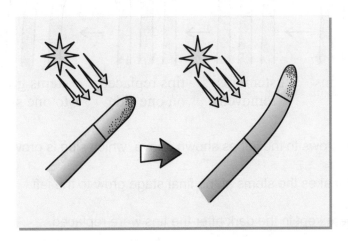

Warm-Up and Exam Questions

Have a go at these questions to see what you know, and what you need to go back and revise again.

1) Explain what the transpiration stream is.
2) Where is meristem tissue found?
3) Explain how rooting powder helps when growing cuttings.
4) What is phototropism?
5) Name the plant hormone that controls phototropism.

Exam Questions

1 The diagram shows a root hair cell.

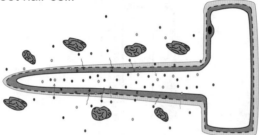

(a) Describe how the structure of a root hair cell is adapted to carry out its function.

(1 mark)

(b) Explain how mineral ions are absorbed into the root hair cell.

(3 marks)

2 In 1918, a Hungarian scientist called Arpad Paal did experiments to investigate how plants grow. The diagram below shows one experiment that he did.

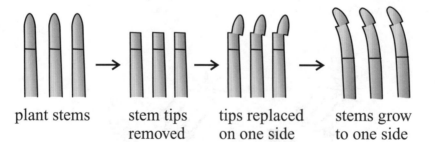

plant stems   stem tips removed   tips replaced on one side   stems grow to one side

(a) When a stem grows to the left as shown above, which side is growing faster?

(1 mark)

(b) Explain what makes the stems at the final stage grow to the left.

(3 marks)

(c) The stems were kept in the dark after the tips were replaced. Explain why.

(1 mark)

Distribution of Organisms

Don't worry — you're nearly at the end of this section. But first you need to get your head around some ecology. These next few pages are all about investigating the distribution of organisms...

The Environment Varies, so Organisms Live in Different Places

1) A habitat is the place where an organism lives, e.g. a playing field.
2) The distribution of an organism is where an organism is found, e.g. in a part of the playing field.
3) Where an organism is found is affected by environmental factors such as:

- Temperature.
- Availability of water.
- Availability of oxygen and carbon dioxide.
- Availability of nutrients.
- Amount of light.

Environmental factors are sometimes called abiotic factors.

4) An organism might be more common in one area than another due to differences in environmental factors between the two areas. For example, in a field, you might find that daisies are more common in the open, than under trees, because there's more light available in the open.
5) There are a couple of ways to study the distribution of an organism. You can:

- measure how common an organism is in two sample areas (e.g. using quadrats) and compare them.
- study how the distribution changes across an area, e.g. by placing quadrats along a transect (p.90).

Use Quadrats to Study The Distribution of Small Organisms

A quadrat is a square frame enclosing a known area, e.g. 1 m². To compare how common an organism is in two sample areas, just follow these simple steps:

1) Place a 1 m² quadrat on the ground at a random point within the first sample area. E.g. divide the area into a grid and use a random number generator to pick coordinates.
2) Count all the organisms within the quadrat.
3) Repeat steps 1 and 2 as many times as you can.
4) Work out the mean number of organisms per quadrat within the first sample area.

A quadrat

- For example, Anna counted the number of daisies in 7 quadrats within her first sample area and recorded the following results: 18, 20, 22, 23, 23, 23, 25
- Here the MEAN is: $\dfrac{\text{TOTAL number of organisms}}{\text{NUMBER of quadrats}} = \dfrac{154}{7} = 22$ daisies per quadrat.
- The MODE is the MOST COMMON value. In this example it's 23.
- And the MEDIAN is the MIDDLE value, when they're in order of size. In this example it's 23 also.

5) Repeat steps 1 to 4 in the second sample area.
6) Finally compare the two means. E.g. you might find 2 daisies per m² in the shade, and 22 daisies per m² (lots more) in the open field.

Count all the organisms in a quadrat, but first remember to...

...put down your quadrat in a random place before you start counting. Even chucking the quadrat over your shoulder is better than putting it down on the first big patch of organisms that you see.

Distribution of Organisms

I'm afraid there's a bit more coming up about the <u>distribution of organisms</u>...

You Can Work Out **Population Size**

To work out the <u>population size</u> of an organism in one sample area:

1) Work out the <u>mean number of organisms per m^2</u>.
   (If your quadrat has an area of 1 m^2, this is the <u>same</u> as the mean
   number of organisms per quadrat, worked out on the previous page.)

2) Then multiply the <u>mean</u> by the <u>total area</u> (in m^2) of the habitat.

> E.g. if the area of an open field is <u>800 m^2</u>, and there are <u>22 daisies per m^2</u>,
> then the size of the daisy population is <u>22 × 800 = 17 600</u>.

Transects Show How Organisms are **Distributed Along a Line**

You can use lines called <u>transects</u> to help find out how organisms (like plants) are <u>distributed</u> across
an area — e.g. if an organism becomes <u>more or less common</u> as you move from a hedge towards the
middle of a field. Here's what to do:

1) <u>Mark out a line</u> in the area you want to study using a tape measure.

2) Then <u>collect data</u> along the line.

3) You can do this by just <u>counting</u> all the organisms you're interested in that <u>touch</u> the line.

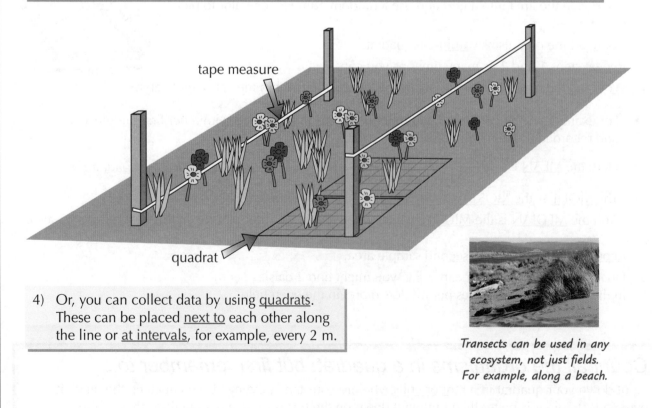

tape measure

quadrat

4) Or, you can collect data by using <u>quadrats</u>.
   These can be placed <u>next to</u> each other along
   the line or <u>at intervals</u>, for example, every 2 m.

*Transects can be used in any
ecosystem, not just fields.
For example, along a beach.*

Distribution of Organisms

When **Collecting Environmental Data** You Need to Think About...

1) Reliability

1) <u>Quadrats</u> and <u>transects</u> are <u>pretty good tools</u> for finding out how an organism is distributed.

2) But, you have to work hard to make sure your results are <u>reliable</u> — which means making sure they are <u>repeatable</u> and <u>reproducible</u> (see page 3).

3) To make your results <u>more</u> reliable you need to:

- Take a <u>large sample size</u>, e.g. use as many quadrats and transects as possible in your sample area. Bigger samples are more representative of the whole population.
- Use <u>random</u> samples, e.g. randomly put down or mark out your quadrat or transect. If all your samples are in <u>one spot</u>, and everywhere else is <u>different</u>, the results you get won't be <u>reproducible</u>.

2) Validity

1) For your results to be <u>valid</u> they must be <u>reliable</u> (see above) and <u>answer the original question</u>.

2) To answer the original question, you need to <u>control all the variables</u>.

3) The question you want to answer is whether a <u>difference in distribution</u> between two sample areas is <u>due</u> to a <u>difference in one environmental factor</u>.

4) If you've controlled all the <u>other variables</u> that could be affecting the distribution, you'll know whether a <u>difference in distribution</u> is caused by the <u>environmental factor</u> or not.

5) If you <u>don't</u> control the other variables you <u>won't know</u> whether any correlation you've found is because of <u>chance</u>, because of the <u>environmental factor</u> you're looking at or because of a <u>different variable</u> — the study <u>won't</u> give you <u>valid data</u>.

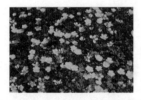

Look back at page 89 for the different environmental factors that can affect distribution. E.g. some types of buttercups are more common in moist areas.

Your investigation needs to be reliable and valid

Take a look back at <u>How Science Works</u> (page 3) to remind yourself all about <u>reliability</u> and <u>validity</u>. If you need to make an investigation using <u>quadrats</u> or <u>transects</u> more reliable, start by thinking about the <u>sample size</u> and how to take samples <u>randomly</u>. The results will be meaningless otherwise.

Warm-Up and Exam Questions

It's finally the end of the section, but before you go on to the next one a few questions need answering.

Warm-Up Questions

1) What do we mean by the 'distribution of an organism'?
2) What is a quadrat?
3) Claire used a 1 m² quadrat to count the number of buttercups in a sample area.
   She placed the quadrat nine times. Here are her results: 2, 15, 4, 6, 8, 3, 11, 10, 9.
   Work out the median number of buttercups per quadrat in her sample area.
4) What is the name of a line used to measure the distribution of organisms across an area?

Exam Questions

1   Paul investigated the distribution of dandelions.
    He counted the number of dandelions in 10 quadrats in five different fields.
    His quadrat measured 1 m². Paul's results are shown in the table below.

Field	Mean number of dandelions per quadrat
A	10
B	35
C	21
D	37
E	21

(a)   What is the mode of Paul's data?

(1 mark)

(b) (i) A week later Paul repeated his experiment in a sixth field, Field F.
         His results for each quadrat are shown below:

         6   15   9   14   20   5   3   11   10   7

         Using this data, estimate the mean number of dandelions per m² in Field F.

(2 marks)

(ii) Field F measures 90 m by 120 m.
      Estimate the number of dandelions in the whole of Field F.

(2 marks)

(c)   Paul's friend Anna also investigated the number of dandelions
        in the same six fields. Anna also used a 1 m² quadrat.
        She counted the number of dandelions in 20 quadrats per field.

        Which investigation, Paul's or Anna's, is likely to produce more reliable data?
        Give a reason for your answer.

(1 mark)

Revision Summary for Section 3

Believe it or not, it's already time for another round of questions. Do as many as you can and if there are some that you're finding really fiddly, don't panic. Have a quick flick over the relevant topics and give the questions another go once you've had another chance to read the pages. Good luck — not that you need it.

1) Write down the word equation for photosynthesis.

2) What is the green substance in leaves that absorbs sunlight?

3) Explain why it's important that a plant doesn't get too hot.

4) Describe three things that a gardener could do to make sure she grows a good crop of tomatoes in her greenhouse.

5) Why is glucose turned into starch when plants need to store it for later?

6) Write down four other ways that plants can use the glucose produced by photosynthesis.

7) What is the advantage to a plant of having root hairs?

8) What name is given to the parts of plants where mitotically active cells are found?

9) Name two types of tissue that the unspecialised cells in plants can turn into.

10) What is a cutting?

11) What do cuttings grow into?

12) Why are cuttings useful?

13) What can be added to soil to encourage cuttings to grow roots?

14) Are shoots negatively or positively phototropic?

15) Explain how auxins cause plant shoots to grow towards light.

16) What is a habitat?

17) Give five environmental factors that can affect the distribution of organisms.

18) Briefly describe how you could find out how common an organism is in two sample areas using quadrats.

19) Describe one way of using a transect to find out how an organism is distributed across an area.

Atoms and Compounds

Atoms contain three types of particle — protons, neutrons and electrons.

Atomic Number and Mass Number Describe an Atom

These two numbers tell you how many of each kind of particle an atom has.

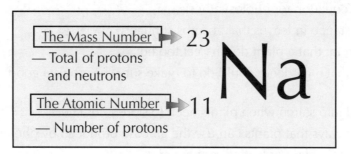

The Mass Number → 23
— Total of protons
and neutrons

Na

The Atomic Number → 11
— Number of protons

1) The atomic number tells you how many protons there are.

2) Atoms of the same element all have the same number of protons —
so atoms of different elements will have different numbers of protons.

3) To get the number of neutrons, just subtract the atomic number from the mass number.
Electrons aren't counted in the mass number because their relative mass is very small.

Particle	Relative Mass	Charge
Proton	1	+1
Neutron	1	0
Electron	0.0005	−1

Electron mass is often
taken as zero.

- Protons are heavy and positively charged
- Neutrons are heavy and neutral
- Electrons are tiny and negatively charged

Compounds are Chemically Bonded

1) Compounds are formed when atoms of two or more elements are chemically combined together.
For example, carbon dioxide is a compound formed from a chemical reaction
between carbon and oxygen.

2) It's difficult to separate the two original elements out again.

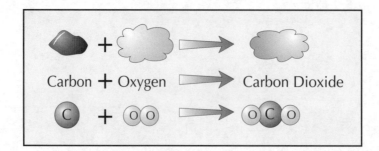

Carbon + Oxygen → Carbon Dioxide

Atomic number = number of protons

If you look at a periodic table you can find the atomic number and mass number of any element.
You can then use these to find out the number of protons, neutrons or electrons in that element.

Isotopes

This page is all to do with the stuff inside the nucleus...

Isotopes are the Same Except for an Extra **Neutron** or Two

Isotopes are: different atomic forms of the same element, which have the SAME number of PROTONS but a DIFFERENT number of NEUTRONS.

1) The upshot is: isotopes must have the same atomic number but different mass numbers.

2) If they had different atomic numbers, they'd be different elements altogether.

3) Carbon-12 and carbon-14 are a very popular pair of isotopes.

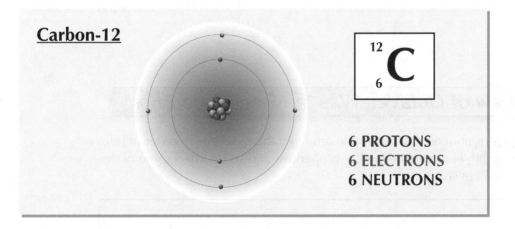

Carbon-12

$^{12}_{6}\text{C}$

6 PROTONS
6 ELECTRONS
6 NEUTRONS

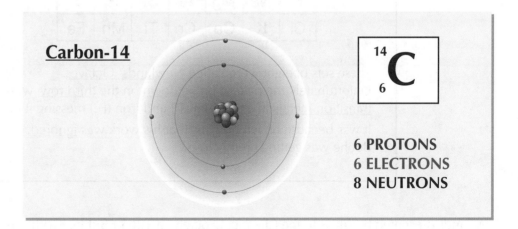

Carbon-14

$^{14}_{6}\text{C}$

6 PROTONS
6 ELECTRONS
8 NEUTRONS

Will you pass your exams — isotope so...

Carbon-14 is unstable. It makes up about one ten-millionth of the carbon in living things. When things die, the C-14 is trapped inside the dead material, and it gradually decays into nitrogen. So by measuring the proportion of C-14 found in some old wood you can calculate how long ago it was living wood.

History of the Periodic Table

We haven't always known as much about Chemistry as we do now. Early chemists looked to try and understand <u>patterns</u> in the elements' properties to get a bit of understanding.

Döbereiner Tried to Organise Elements into Triads

1) Back in the 1800s the only thing they could measure was <u>relative atomic mass</u> (see page 125), and so the <u>known</u> elements were arranged <u>in order of atomic mass</u>.

2) In 1828 a guy called <u>Döbereiner</u> started to put this list of elements into groups based on their <u>chemical properties</u>. He put the elements into groups of <u>three</u>, which he called <u>triads</u>. E.g. Cl, Br and I were one triad, and Li, Na and K were another.

3) The <u>middle element</u> of each triad had a relative atomic mass that was the <u>average</u> of the other two.

Element	Relative atomic mass
Lithium	7
Sodium	23
Potassium	39

(7 + 39) ÷ 2 = 23

Newlands' Law of Octaves Was the First Good Effort

A chap called <u>Newlands</u> noticed that when you arranged the elements in order of relative atomic mass, every <u>eighth</u> element had similar properties, and so he listed some of the known elements in rows of seven:

H	Li	Be	B	C	N	O
F	Na	Mg	Al	Si	P	S
Cl	K	Ca	Cr	Ti	Mn	Fe

These sets of eight were called <u>Newlands' Octaves</u>.
Unfortunately the pattern <u>broke down</u> on the <u>third row</u>, with <u>transition metals</u> like titanium (Ti) and iron (Fe) messing it up.

It was because he left <u>no gaps</u> that his work was <u>ignored</u>.
But he was getting <u>pretty close</u>.

Newlands presented his ideas to the Chemical Society in 1865. But his work was criticised because:

1) His groups contained elements that didn't have <u>similar properties</u>, e.g. <u>carbon</u> and <u>titanium</u>.

2) He <u>mixed up metals and non-metals</u> e.g. <u>oxygen</u> and <u>iron</u>.

3) He <u>didn't leave any gaps</u> for elements that hadn't been discovered yet.

History of the Periodic Table

Newlands wasn't the only one who had ideas about classifying elements.

Dmitri Mendeleev Left Gaps and Predicted New Elements

1) In 1869, Dmitri Mendeleev in Russia, armed with about 50 known elements, arranged them into his Table of Elements — with various gaps as shown.

```
              Mendeleev's Table of the Elements
   H
   Li  Be                                    B  C  N  O  F
   Na  Mg                                    Al Si P  S  Cl
   K   Ca * Ti V  Cr Mn Fe Co Ni Cu Zn *  *  As Se Br
   Rb  Sr  Y Zr Nb Mo *  Ru Rh Pd Ag Cd In Sn Sb Te I
   Cs  Ba  * * Ta W  *  Os Ir Pt Au Hg Tl Pb Bi
```

2) Mendeleev put the elements in order of atomic mass (like Newlands did).

3) But Mendeleev found he had to leave gaps in order to keep elements with similar properties in the same vertical groups — and he was prepared to leave some very big gaps in the first two rows before the transition metals come in on the third row.

4) The gaps were the really clever bit because they predicted the properties of so far undiscovered elements.

5) When they were found and they fitted the pattern it helped confirm Mendeleev's ideas. For example, Mendeleev made really good predictions about the chemical and physical properties of an element he called ekasilicon, which we know today as germanium.

Not All Scientists Thought the Periodic Table was Important

1) When the periodic table was first released, many scientists thought it was just a bit of fun.

2) At that time, there wasn't all that much evidence to suggest that the elements really did fit together in that way.

3) After Mendeleev released his work, newly discovered elements fitted into the gaps he left.

4) This was convincing evidence in favour of the periodic table.

Elementary my dear Mendeleev

Mendeleev tried to classify the 50 elements that were known at the time — and although his attempt wasn't perfect, he got on pretty well. It was an improvement on Newlands' attempt anyway.

98

The Modern Periodic Table

Chemists were getting pretty close to producing something useful.
The big breakthrough came when the structure of the atom was understood a bit better.

The Periodic Table Separates the Metals and Non-metals

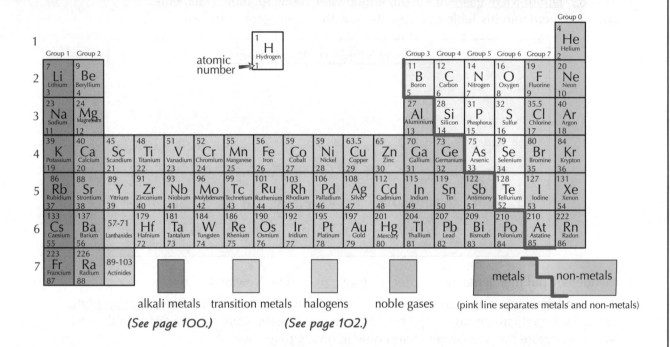

alkali metals   transition metals   halogens   noble gases   (pink line separates metals and non-metals)

(See page 100.)   *(See page 102.)*

The Periodic Table is Arranged in Periods and Groups

Periods

1) The rows of the periodic table are called periods.

2) The elements are arranged in order of increasing atomic number along each row. E.g. the atomic numbers of the Period 2 elements increase from 3 for Li to 10 for Ne.

3) The period number is the same as the number of electron shells. E.g. the Period 3 elements have 3 electron shells.

4) The properties of elements change as you go along a period (sometimes quite dramatically).

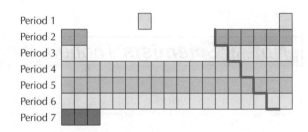

Groups

1) The columns of the periodic table are called groups.

2) Elements in the same group have similar properties. This is because they have the same number of electrons in their outer shell.

3) The group number is always equal to the number of electrons in the outer shell.
E.g. the Group 2 elements all have two electrons in their outer shell.

4) The properties of elements (such as reactivity) often gradually change as you go down a group (as the atomic number increases).

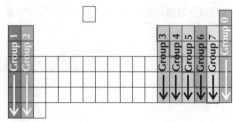

SECTION 4 — THE PERIODIC TABLE, BONDING AND CALCULATIONS

Warm-Up and Exam Questions

These questions will help you find out if you've learnt all the basics about atoms, compounds, isotopes and the periodic table. Have a look back through the last few pages if you're unsure about any of these questions. It's really important to get these basics right.

Warm-Up Questions

1) Explain the difference between mass number and atomic number.
2) What were Döbereiner's triads?
3) In the modern periodic table, give one thing that you can tell about an element from its group number.
4) What are the rows of the periodic table called?

Exam Questions

1 Carbon has several isotopes, for example carbon-12 and carbon-13.
  Details about the carbon-13 isotope are shown below:

$$^{13}_{6}C$$

  (a) Explain what isotopes are.

  (3 marks)

  (b) Draw a diagram to represent the carbon-13 atom. Label the number of protons and neutrons in the nucleus and show the electron arrangement.

  (3 marks)

2 (a) A proton has a relative mass of 1. What is the relative mass of a neutron?

  (1 mark)

  (b) Electrons aren't counted in the mass number of elements.
      Give a reason for this.

  (1 mark)

3 Which of the following statements about Mendeleev's periodic table is **not** true?

  A Elements with similar chemical properties were placed in vertical groups.

  B Gaps were left which helped in predicting the properties of undiscovered elements.

  C The elements were arranged in order of atomic mass.

  D Mendeleev's periodic table contained over 100 elements.

  (1 mark)

4 Newlands and Mendeleev both came up with a system for classifying elements.

  (a) Describe Newlands' system for classifying elements.

  (1 mark)

  (b) Explain why the work of Mendeleev was taken more seriously than that of Newlands.

  (4 marks)

Group 1 — The Alkali Metals

The alkali metals are <u>silvery solids</u> that have to be <u>stored in oil</u> and handled with <u>forceps</u> (they burn the skin).

Group 1 Metals are Known as the 'Alkali Metals'

Group 1 metals include lithium, sodium and potassium...

> As you go <u>DOWN</u> Group 1, the alkali metals become <u>more reactive</u> — the <u>outer electron</u> is more easily <u>lost</u>, because it's further from the nucleus (the <u>atomic radius</u> is <u>larger</u>) so <u>less energy</u> is needed to remove it.

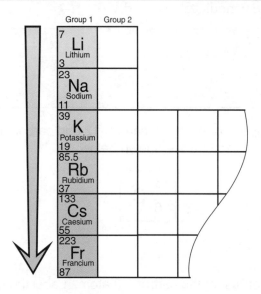

The alkali metals all have <u>ONE outer electron</u>.
This makes them <u>very reactive</u> and gives them all <u>similar properties</u>.
They all have the following <u>physical properties</u>:

- <u>Low melting point</u> and <u>boiling point</u> (compared with other metals).
- <u>Low density</u> — lithium, sodium and potassium float on water.
- <u>Very soft</u> — they can be cut with a knife.

The alkali metals always form <u>ionic</u> compounds (page 106). They are so keen to lose the outer electron there's no way they'd consider <u>sharing</u>, so covalent bonding (page 112) is <u>out of the question</u>.

Oxidation is the Loss of Electrons

Group 1 metals are keen to <u>lose an electron</u> to form a <u>1⁺ion</u> with a <u>stable electronic structure</u>.

The <u>more</u> reactive the metal the happier it is to <u>lose</u> an electron.

Loss of electrons is called <u>OXIDATION</u>.

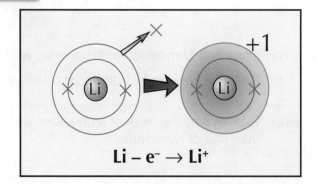

$$Li - e^- \rightarrow Li^+$$

Group 1 — The Alkali Metals

This page covers loads more interesting facts about the <u>alkali metals</u>. I knew you'd be excited.

Reaction with **Cold Water** *Produces* **Hydrogen Gas**

1) When <u>lithium</u>, <u>sodium</u> or <u>potassium</u> are put in <u>water</u>, they react very <u>vigorously</u>.
2) They <u>move</u> around the surface, <u>fizzing</u> furiously.
3) They produce <u>hydrogen</u>.
4) Potassium gets hot enough to <u>ignite</u> it.
5) If it hasn't already been ignited by the reaction, a lighted splint will <u>indicate</u> hydrogen by producing the notorious "<u>squeaky pop</u>" as it ignites.

Squeaky pop

6) The reaction makes an <u>alkaline solution</u> — this is why Group 1 is known as the <u>alkali</u> metals.
7) A <u>hydroxide</u> of the metal forms, e.g. sodium hydroxide (NaOH), potassium hydroxide (KOH) or lithium hydroxide (LiOH).

$$2Na_{(s)} + 2H_2O_{(l)} \rightarrow 2NaOH_{(aq)} + H_2{(g)}$$
$$2K_{(s)} + 2H_2O_{(l)} \rightarrow 2KOH_{(aq)} + H_2{(g)}$$

8) This experiment shows the <u>relative reactivities</u> of the alkali metals.
9) The <u>more violent</u> the reaction, the <u>more reactive</u> the alkali metal is.

Reaction with **Chlorine** *Produces* **Salts**

1) Alkali metals react vigorously with <u>chlorine</u>.
2) The reaction produces <u>colourless crystalline salts</u>, e.g. lithium chloride (LiCl), sodium chloride (NaCl) and potassium chloride (KCl).

$$2Na_{(s)} + Cl_2{(g)} \rightarrow 2NaCl_{(s)}$$
$$2K_{(s)} + Cl_2{(g)} \rightarrow 2KCl_{(s)}$$

That reaction with water is the reason they're called alkali metals

The alkali metals all have very similar properties due to the fact that they've all got the same number of electrons in their outer shell. This means that they all react in the same way, for example, with water and with chlorine. Once you know about one alkali metal, you know about them all.

Group 7 — The Halogens

The 'trend thing' happens in <u>Group 7</u> as well — that shouldn't come as a surprise.
But some of the trends are kind of the opposite of the Group 1 trends.

Group 7 Elements are Known as the 'Halogens'

1) Group 7 is made up of fluorine, chlorine, bromine, iodine and astatine.

2) All Group 7 elements have <u>7 electrons in their outer shell</u> — so they all react
   by <u>gaining one electron</u> to form a negative ion.

3) This means they've all got <u>similar properties</u>.

As you go <u>DOWN</u> Group 7, the halogens become <u>less reactive</u> —
there's less inclination to gain the <u>extra electron</u> to fill the outer shell
when it's <u>further out</u> from the nucleus (there's a <u>larger atomic radius</u>).

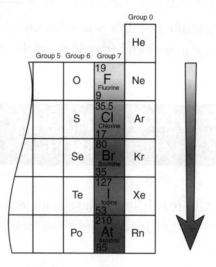

4) As you go <u>down group 7</u> the <u>melting points</u> and <u>boiling points</u> of the halogens <u>increase</u>.

5) This means that at <u>room temperature</u>:

- <u>Chlorine</u> (Cl$_2$) is a fairly reactive, poisonous, <u>dense green gas</u> (low boiling point).
- <u>Bromine</u> (Br$_2$) is a dense, poisonous, <u>orange liquid</u>.
- <u>Iodine</u> (I$_2$) is a <u>dark grey crystalline solid</u> (high boiling point).

Reduction is the Gain of Electrons

1) Halogens are keen to <u>gain an electron</u> to form
   a <u>1⁻ ion</u> with a <u>stable electronic structure</u>.

2) The <u>more</u> reactive the halogen the happier it is
   to <u>gain</u> an electron.

3) Gain of electrons is called <u>REDUCTION</u>.

$$Cl_2 + 2e^- \rightarrow 2Cl^-$$

| Halogen molecule | Halide ion |

Group 7 — The Halogens

Now that you've learnt the basics about the halogens it's time to spice things up with some reactions...

Halogens become Less Reactive Down the Group

Chlorine is more reactive than bromine, which is more reactive than iodine.
The trend in reactivity of the halogens can be shown by looking at different reactions.

Reactions with Alkali Metals

1) The halogens react with alkali metals like lithium, sodium and potassium to form salts called metal halides e.g. sodium chloride (NaCl), potassium bromide (KBr) and lithium iodide (LiI).
2) The reactions become less vigorous as you go down the group.

$$2Na_{(s)} + Cl_{2(g)} \rightarrow 2NaCl_{(s)}$$
$$2K_{(s)} + Br_{2(g)} \rightarrow 2KBr_{(s)}$$

Reactions with Iron

1) They react with iron to form coloured solids called iron halides.
2) Again, the reactions become less vigorous as you go down the group.

$$2Fe_{(s)} + 3Cl_{2(g)} \rightarrow 2FeCl_{3(s)}$$
$$2Fe_{(s)} + 3Br_{2(g)} \rightarrow 2FeBr_{3(s)}$$

Displacement Reactions

These displacement reactions can be used to determine the relative reactivity of the halogens.

1) A displacement reaction is where a more reactive element 'pushes out' (displaces) a less reactive element from a compound.
2) Chlorine is more reactive than iodine. So chlorine reacts with potassium iodide solution to form potassium chloride, and the iodine is left in solution.
3) Chlorine can also displace bromine from solutions of bromides.
4) Bromine will displace iodine because of the trend in reactivity.

$$Cl_{2(g)} + 2KI_{(aq)} \rightarrow I_{2(aq)} + 2KCl_{(aq)}$$
$$Cl_{2(g)} + 2KBr_{(aq)} \rightarrow Br_{2(aq)} + 2KCl_{(aq)}$$

Polish that halo and get revising...

Displacement reactions needn't be too confusing — they're just all about figuring out what's more reactive than what. If a halogen is higher up the group than another one, it'll kick that one out of the solution and get in there itself. Just like how I displace my sister from the sofa before Corrie.

Warm-Up and Exam Questions

These questions are all about the groups of the periodic table that you need to know about.
Treat the exam questions like the real thing — don't look back through the book until you've finished.

Warm-Up Questions

1) In Group 1, as you go down the periodic table, does the reactivity increase or decrease?
2) Which gas is produced when an alkali metal reacts with water?
3) Give an example of a salt produced when a Group 1 metal reacts with a Group 7 element.
4) Write down the balanced equation for the reaction between iron and chlorine.
5) Write down the balanced equation for the displacement of bromine from potassium bromide by chlorine.

Exam Questions

1   The table shows some of the physical properties of three of the halogens.

	Properties			
Halogen	Atomic number	Colour	Physical state at room temperature	Boiling point
Chlorine	17	green		−34 °C
Bromine	35	red-brown		59 °C
Iodine	53	dark grey		185 °C

(a) Complete the table to give the physical state at room temperature
of all three halogens.

(3 marks)

(b) Draw an arrow next to the left hand side of the table to show the direction of
increasing reactivity in the halogens.

(1 mark)

(c) This equation shows a reaction between chlorine and potassium iodide.

$$Cl_{2(g)} + 2KI_{(aq)} \rightarrow I_{2(aq)} + 2KCl_{(aq)}$$

(i)   What type of reaction is this?

(1 mark)

(ii) Which is the less reactive halogen in this reaction?

(1 mark)

2   All the Group 1 metals react vigorously with water.
(a) Explain why this is.

(1 mark)

(b) The Group 1 metals all react with water at a different rate. Explain why this is.

(2 marks)

(c) Give the two products formed in these reactions.

(2 marks)

(d) During these reactions a solution is formed.
Is this solution acidic, neutral or alkaline?

(1 mark)

Ionic Bonding

Ionic bonding is one of the ways atoms can form <u>compounds</u>.

Ionic Bonding — Transferring Electrons

In <u>ionic bonding</u>, atoms <u>lose or gain electrons</u> to form <u>charged particles</u> (called <u>ions</u>) which are then <u>strongly attracted</u> to one another (because of the attraction of opposite charges, + and –).

A shell with just one electron is well keen to get rid...

1) <u>All</u> the atoms over at the <u>left-hand side</u> of the periodic table, e.g. <u>sodium, potassium, calcium</u> etc. have just <u>one or two electrons</u> in their outer shell (highest energy level).

2) And they're <u>pretty keen to get shot of them</u>, because then they'll only have <u>full shells</u> left, which is how they <u>like</u> it. (They try to have the same <u>electronic structure</u> as a <u>noble gas</u>.)

3) So given half a chance they do get rid, and that leaves the atom as an <u>ion</u> instead.

4) Now ions aren't the kind of things that sit around quietly watching the world go by. They tend to <u>leap</u> at the first passing ion with an <u>opposite charge</u> and stick to it like glue.

A nearly full shell is well keen to get that extra electron...

1) On the <u>other side</u> of the periodic table, the elements in <u>Group 6</u> and <u>Group 7</u>, such as <u>oxygen</u> and <u>chlorine</u>, have outer shells which are <u>nearly full</u>.

2) They're obviously pretty keen to <u>gain</u> that <u>extra one or two electrons</u> to fill the shell up.

3) When they do, of course, they become <u>ions</u> and before you know it, <u>pop</u>, they've latched onto the atom (ion) that gave up the electron a moment earlier.

The reaction of sodium and chlorine is a <u>classic case</u>:

The <u>sodium</u> atom <u>gives up</u> its <u>outer electron</u> and becomes an Na^+ ion.

The <u>chlorine</u> atom has <u>picked up</u> the <u>spare electron</u> and becomes a Cl^- ion.

POP!

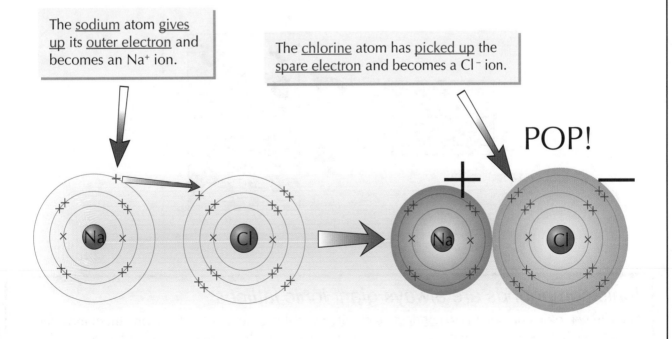

Ionic Bonding

Ionic bonds produce giant ionic structures.

Ionic Compounds have a Regular Lattice Structure

1) Ionic compounds always have giant ionic lattices.

2) The ions form a closely packed regular lattice arrangement.

3) There are very strong electrostatic forces of attraction between oppositely charged ions, in all directions.

4) A single crystal of sodium chloride (salt) is one giant ionic lattice, which is why salt crystals tend to be cuboid in shape. The Na^+ and Cl^- ions are held together in a regular lattice.

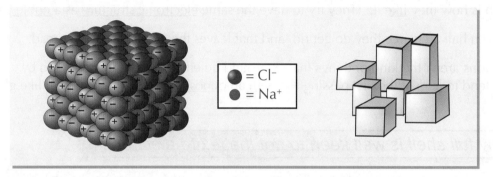

Ionic Compounds all have Similar Properties

1) They all have high melting points and high boiling points due to the strong attraction between the ions. It takes a large amount of energy to overcome this attraction. When ionic compounds melt, the ions are free to move and they'll carry electric current.

2) They do dissolve easily in water though. The ions separate and are all free to move in the solution, so they'll carry electric current.

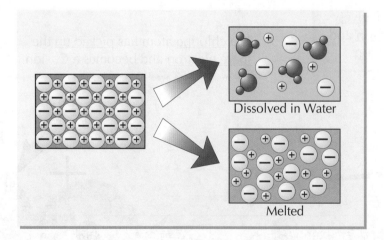

Dissolved in Water

Melted

Ionic compounds are always giant ionic lattices

You can get ionic compounds to conduct electricity by melting them or by dissolving them in water. Dissolving them is easier though as it takes a lot of energy to melt an ionic compound. Either way, it's the free ions that carry the electric current.

Ions

Make sure you've really got your head around the idea of ionic bonding before you start on this page.

Groups *1 & 2* and *6 & 7* are the Most Likely to Form *Ions*

1) Remember, atoms that have <u>lost</u> or <u>gained</u> an electron (or electrons) are <u>ions</u>.

2) Ions have the <u>electronic structure</u> of a <u>noble gas</u>.

3) The elements that most readily form ions are those in <u>Groups 1</u>, <u>2</u>, <u>6 and 7</u>.

4) <u>Group 1 and 2 elements</u> are <u>metals</u> and they <u>lose</u> electrons to form <u>positive ions</u>.

5) For example, <u>Group 1</u> elements (the <u>alkali metals</u>) form ionic compounds with <u>non-metals</u> where the metal ion has a 1^+ charge. E.g. K^+Cl^-.

6) <u>Group 6 and 7 elements</u> are <u>non-metals</u>. They <u>gain</u> electrons to form <u>negative ions</u>.

7) For example, <u>Group 7</u> elements (the <u>halogens</u>) form ionic compounds with the <u>alkali metals</u> where the halide ion has a 1^- charge. E.g. Na^+Cl^-.

8) The <u>charge</u> on the <u>positive ions</u> is the <u>same</u> as the <u>group number</u> of the element:

POSITIVE IONS		NEGATIVE IONS	
Group 1	Group 2	Group 6	Group 7
Li^+ Na^+ K^+	Be^{2+} Mg^{2+} Ca^{2+}	O^{2-}	F^- Cl^-

9) Any of the positive ions above can <u>combine</u> with any of the negative ions to form an <u>ionic compound</u>.

10) Only elements at <u>opposite sides</u> of the periodic table will form ionic compounds, e.g. Na and Cl, where one of them becomes a <u>positive ion</u> and one becomes a <u>negative ion</u>.

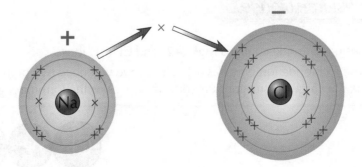

Remember, the + and – charges we talk about, e.g. Na^+ for sodium, just tell you <u>what type of ion the atom WILL FORM</u> in a chemical reaction. In sodium <u>metal</u> there are <u>only neutral sodium atoms, Na</u>. The Na^+ ions <u>will only appear</u> if the sodium metal <u>reacts</u> with something like water or chlorine.

Formulas of Ionic Compounds

This page will help you work out the chemical formula of an ionic compound.

Look at **Charges** to Work out the **Formula** of an **Ionic Compound**

1) Ionic compounds are made up of a positively charged part and a negatively charged part.

2) The overall charge of any compound is zero.

3) So all the negative charges in the compound must balance all the positive charges.

4) You can use the charges on the individual ions present to work out the formula for the ionic compound:

Sodium chloride

Sodium chloride contains Na^+ (+1) and Cl^- (–1) ions.

(+1) + (–1) = 0. The charges are balanced with one of each ion, so the formula for sodium chloride = NaCl.

NaCl

Magnesium chloride

Magnesium chloride contains Mg^{2+} (+2) and Cl^- (–1) ions.

Because a chloride ion only has a 1^- charge we will need two of them to balance out the 2^+ charge of a magnesium ion. This gives us the formula $MgCl_2$.

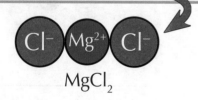

$MgCl_2$

The formula for exam success = revision...

The main thing to remember is that in compounds the total charge must always add up to zero.
So, for example, if you've got 2 positive charges, you need to balance it out with 2 negative charges.
Easy as that — as long as you know the charges of the ions involved, you can work out any formula.

Electronic Structure of Ions

This page has some lovely drawings of the <u>electronic structures</u> of ions.
As well as being pretty, they're really useful too.

*Show the **Electronic Structure** of **Simple** Ions with **Diagrams***

A useful way of representing ions is by <u>drawing</u> out their electronic structure. Just use a big <u>square bracket</u> and a + or − to show the charge. A few <u>ions</u> and the <u>ionic compounds</u> they form are shown below:

Sodium Chloride

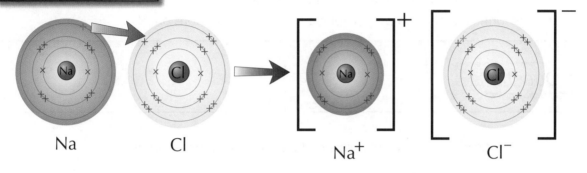

NaCl (Sodium Chloride)

Magnesium Oxide

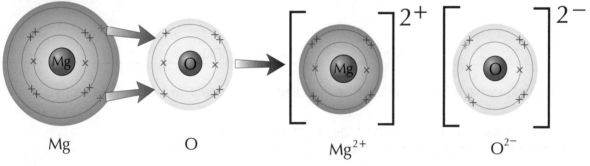

MgO (Magnesium Oxide)

Calcium Chloride

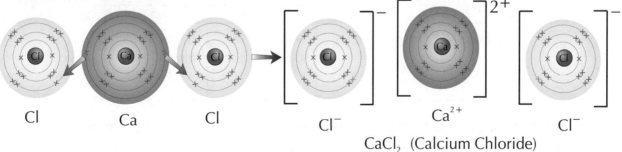

CaCl₂ (Calcium Chloride)

Show the electronic structure of ions with square brackets

Whether or not you're able to reproduce the drawings on this page all comes down to how well you've understood <u>ionic bonding</u>. (So if you're struggling, try reading the last few pages again.)

Warm-Up and Exam Questions

These questions will help you find out if you've learnt all the basics about ions and ionic bonding. Have a look back through the last few pages if you're unsure about any of these questions. It's tempting just to skip past anything you don't know, but it won't help you in the exam.

Warm-Up Questions

1) Sodium chloride has a giant ionic structure. Does it have a high or a low boiling point?
2) Why do ionic compounds conduct electricity when dissolved?
3) Do elements from Group 1 form positive ions or negative ions?
4) Do elements from Group 7 form positive ions or negative ions?
5) What is the formula of the compound containing Al^{3+} and OH^- ions only?

Exam Questions

1   The diagrams below show the electronic structures of sodium and fluorine.

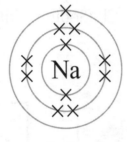

Sodium

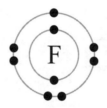

Fluorine

(a)   Describe what will happen when sodium and fluorine react, in terms of electrons.

(2 marks)

(b)   When sodium and fluorine react they form an ionic compound
      Describe the structure of an ionic compound.

(3 marks)

2   The table below lists different ions.

Positive Ions		Negative Ions	
1^+ ions	2^+ ions	2^- ions	1^- ions
Lithium Li^+	Magnesium Mg^{2+}	Carbonate CO_3^{2-}	Hydroxide OH^-
		Sulfate SO_4^{2-}	Nitrate NO_3^-

Use the information in the table to write the formulas of the following ionic compounds:

(a)   magnesium carbonate

(1 mark)

(b)   lithium sulfate

(1 mark)

3   When lithium reacts with oxygen it forms an ionic compound, Li_2O.
   (a)   Name the compound formed.

   (1 mark)

   (b)   (i)   Complete the left hand side of the diagram below using arrows to show how
               the electrons are transferred when Li_2O is formed.

   (1 mark)

        (ii)   Show the electron arrangements and the charges on the ions formed.

   (2 marks)

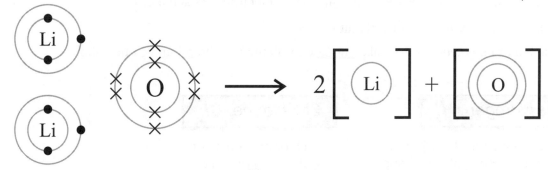

4   Magnesium (atomic number 12) and fluorine (atomic number 9) combine vigorously to form
     magnesium fluoride, an ionic compound.
   (a)   Draw a diagram to show the electron arrangement of each atom.

   (2 marks)

   (b)   Give the symbol (including the charge) for each of the ions formed.

   (2 marks)

   (c)   Using your answer to part (b), work out the formula of magnesium fluoride.

   (1 mark)

   (d)   Once formed, explain why the ions remain together in a compound.

   (1 mark)

   (e)   Magnesium fluoride has a giant ionic structure. Explain why:
        (i)   it doesn't melt easily.

   (2 marks)

        (ii)   it conducts electricity when molten.

   (1 mark)

5   Potassium and chlorine react to form potassium chloride.
   (a)   Complete the following table.

   (2 marks)

	Potassium atom, K	Potassium ion, K^+	Chlorine atom, Cl	Chloride ion, Cl^-
Number of electrons	19			

   (b)   Draw a diagram to show the formation of potassium chloride.

   (2 marks)

Covalent Bonding

Some elements bond ionically (see page 105) but others form strong <u>covalent bonds</u>.
This is where atoms <u>share electrons</u> with each other so that they've got <u>full outer shells</u>.

Covalent Bonds — Sharing Electrons

1) Sometimes atoms prefer to make <u>covalent bonds</u> by <u>sharing</u> electrons with other atoms.

2) They only share electrons in their <u>outer shells</u> (highest energy levels).

3) This way <u>both</u> atoms feel that they have a <u>full outer shell</u>, and that makes them happy.
Having a full outer shell gives them the electronic structure of a <u>noble gas</u>.

4) Each <u>covalent bond</u> provides one <u>extra</u> shared electron for each atom.

5) So, a covalent bond is a <u>shared pair</u> of electrons.

6) Each atom involved has to make <u>enough</u> covalent bonds to <u>fill up</u> its outer shell.

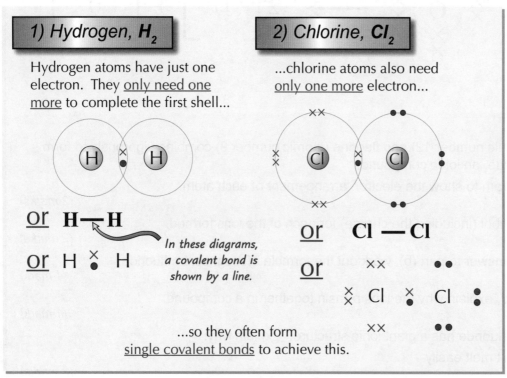

1) Hydrogen, H_2

Hydrogen atoms have just one electron. They <u>only need one more</u> to complete the first shell...

or **H—H**

or H ×• H

In these diagrams, a covalent bond is shown by a line.

2) Chlorine, Cl_2

...chlorine atoms also need <u>only one more</u> electron...

or **Cl — Cl**

or

In a dot and cross diagram, you only have to draw the outer shell of electrons.

...so they often form <u>single covalent bonds</u> to achieve this.

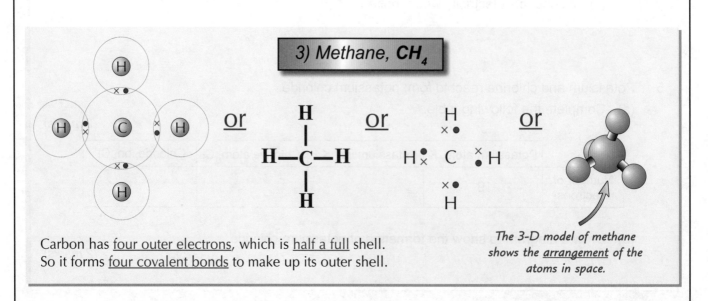

3) Methane, CH_4

or

H—C—H

or

or

Carbon has <u>four outer electrons</u>, which is <u>half a full</u> shell.
So it forms <u>four covalent bonds</u> to make up its outer shell.

The 3-D model of methane shows the <u>arrangement</u> of the atoms in space.

More Covalent Bonding

There are four more examples of covalent bonding on this page — just a few diagrams and a smattering of words. What a pleasant page.

4) Hydrogen Chloride, **HCl**

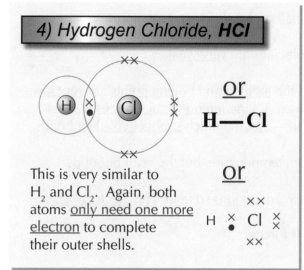

or

H—Cl

or

This is very similar to H_2 and Cl_2. Again, both atoms <u>only need one more electron</u> to complete their outer shells.

5) Ammonia, **NH₃**

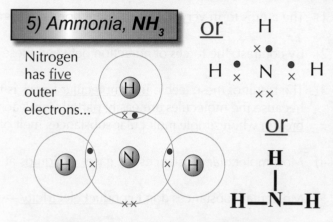

Nitrogen has <u>five</u> outer electrons...

or

...so it needs to form <u>three covalent bonds</u> to make up the extra <u>three</u> electrons needed.

6) Water, **H₂O**

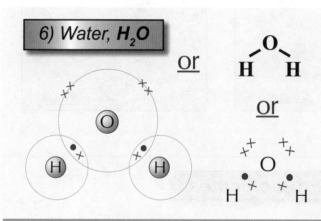

or

or

<u>Oxygen</u> atoms have <u>six</u> outer electrons. They sometimes form <u>ionic</u> bonds by <u>taking</u> two electrons to complete their outer shell. However they'll also cheerfully form <u>covalent bonds</u> and <u>share</u> two electrons instead. In <u>water molecules</u>, the oxygen <u>shares</u> electrons with the two H atoms.

7) Oxygen, **O₂**

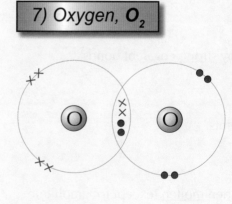

In <u>oxygen gas</u>, oxygen <u>shares two electrons</u> with another oxygen atom to get a full outer shell. A <u>double</u> covalent bond is formed.

or **O=O** or

or

Remember — it's only the outer shells that share electrons with each other.

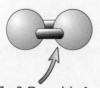

The 3-D model of oxygen shows the <u>atoms</u> and the <u>covalent bonds</u>.

Covalent bonding involves sharing rather than giving electrons

Every atom wants a full outer shell, and they can get that either by becoming an <u>ion</u> (see page 107) or by <u>sharing electrons</u>. Some atoms can share two electron pairs — it's not limited to a single bond.

Covalent Substances

Substances with <u>covalent bonds</u> (electron sharing) can form <u>simple molecules</u> or <u>giant structures</u>.

Simple Molecular Substances

1) The atoms form <u>very strong</u> covalent bonds to form <u>small</u> molecules of several atoms.

2) By contrast, the forces of attraction <u>between</u> these molecules are <u>very weak</u>.

3) The result of these feeble <u>intermolecular forces</u> is that the <u>melting</u> and <u>boiling points</u> are <u>very low</u>, because the molecules are <u>easily parted</u> from each other. It's the <u>intermolecular forces</u> that get <u>broken</u> when simple molecular substances melt or boil — <u>not</u> the much <u>stronger covalent bonds</u>.

4) Most molecular substances are <u>gases or liquids</u> at room temperature, but they can be <u>solids</u>.

5) Molecular substances <u>don't conduct electricity</u> — there are <u>no ions</u> so there's <u>no electrical charge</u>.

Very weak intermolecular forces

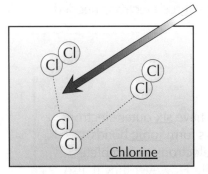

Chlorine

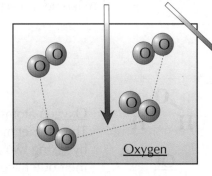

Oxygen

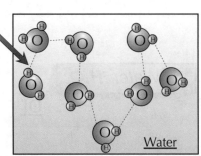

Water

Giant Covalent Structures Are Macromolecules

1) These are similar to giant ionic structures (lattices) <u>except</u> that there are <u>no charged ions</u>.

2) <u>All</u> the atoms are <u>bonded</u> to <u>each other</u> by <u>strong</u> covalent bonds.

3) This means that they have <u>very high</u> melting and boiling points.

4) They <u>don't conduct electricity</u> — not even when <u>molten</u> (except for graphite).

5) The <u>main examples</u> are <u>diamond</u> and <u>graphite</u>, which are both made only from <u>carbon atoms</u>, and <u>silicon dioxide</u> (silica) — see the next page.

Covalent Substances

Here are three examples of <u>giant covalent structures</u>.

Diamond

Each carbon atom forms <u>four covalent bonds</u> in a <u>very rigid</u> giant covalent structure. This structure makes diamond the <u>hardest</u> natural substance, so it's used for drill tips. And it's <u>pretty</u> and <u>sparkly</u> too.

Graphite

Each carbon atom only forms <u>three covalent bonds</u>. This creates <u>layers</u> which are free to <u>slide over each other</u>, like a pack of cards — so graphite is <u>soft</u> and <u>slippery</u>.

The layers are held together so loosely that they can be <u>rubbed off</u> onto paper — that's how a <u>pencil</u> works. This is because there are <u>weak intermolecular forces</u> between the layers.

Graphite is the only <u>non-metal</u> which is a <u>good conductor of heat and electricity</u>. Each carbon atom has one <u>delocalised</u> (free) electron and it's these free electrons that <u>conduct</u> heat and electricity.

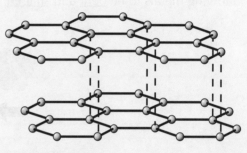

Silicon Dioxide (Silica)

Sometimes called <u>silica</u>, this is what <u>sand</u> is made of. Each grain of sand is <u>one giant structure</u> of silicon and oxygen.

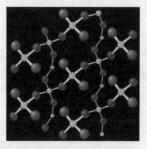

Graphite and diamond contain exactly the same atoms

Graphite and diamond are both made purely from <u>carbon</u> — there's no difference at all in their <u>atoms</u>. The difference in properties (and price) of the two substances is all down to the way the atoms are <u>held together</u>. Don't get confused though — they're both still giant covalent substances.

Metallic Structures

Ever wondered what makes <u>metals</u> tick? Well, either way, this is the page for you.

Metal Properties are All Due to the Sea of Free Electrons

1) <u>Metals</u> also consist of a <u>giant structure</u>.

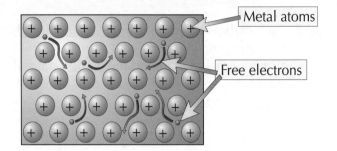

2) <u>Metallic bonds</u> involve the all-important 'free electrons' which produce <u>all</u> the properties of metals. These delocalised (free) electrons come from the <u>outer shell</u> of <u>every</u> metal atom in the structure.

3) These electrons are <u>free to move</u> through the whole structure and so metals are good conductors of <u>heat and electricity</u>.

4) These electrons also <u>hold</u> the <u>atoms</u> together in a <u>regular</u> structure. There are strong forces of <u>electrostatic attraction</u> between the <u>positive metal ions</u> and the <u>negative electrons</u>.

5) They also allow the layers of atoms to <u>slide</u> over each other, allowing metals to be <u>bent</u> and <u>shaped</u>.

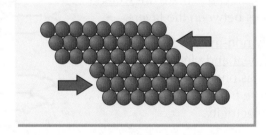

Alloys are Harder than Pure Metals

1) <u>Pure metals</u> often aren't quite right for certain jobs. So scientists <u>mix two or more metals together</u> (or sometimes a metal and a non-metal) — creating an <u>alloy</u> with the properties they want.

2) Different elements have <u>different sized atoms</u>. So when another metal is mixed with a pure metal, the new metal atoms will <u>distort</u> the layers of metal atoms, making it more difficult for them to slide over each other. So alloys are <u>harder</u>.

Identifying Structures

If you know the <u>properties</u> of the four types of substance then this page shouldn't be a problem. If you don't then it's worth taking another look at the last few pages.

Identifying the Structure of a Substance by its Properties

You should be able to easily <u>identify</u> most substances just by the way they <u>behave</u> as either:

- <u>giant ionic</u>,
- <u>simple molecular</u>,
- <u>giant covalent</u>,
- or <u>giant metallic</u>.

You can tell what <u>type</u> of <u>structure</u> a substance has from its <u>physical properties</u>.
Try this one:

> <u>Example</u>:
>
> Four substances were tested for various properties with the following results.
>
> Identify the structure of each substance. (Answers on page 273.)

Substance	Melting point (°C)	Boiling point (°C)	Good electrical conductor?
A	−219.62	−188.12	No
B	1535	2750	Yes
C	1410	2355	No
D	770	1420	When molten

Look at the properties to identify the structure

If you know the properties of a substance then you can work out its structure. It's not as hard as it sounds because a substance will always be one of four things — giant ionic, simple molecular, giant covalent or giant metallic. So, if you know how to identify those four substances, then you know how to identify anything. Phew.

Warm-Up and Exam Questions

Don't charge past this page, it's a lot more important than it looks.

Warm-Up Questions

1) Describe what a covalent bond is.
2) Why does chlorine have a very low boiling point?
3) Describe the differences in the physical properties of diamond and graphite.
4) Give another example of a substance that has a giant covalent structure.

Exam Questions

1   Methane is a covalently bonded molecule with the formula CH_4.
    Draw a dot and cross diagram of a methane molecule,
    showing only the outer electrons.

    (2 marks)

2   The table compares some physical properties of silicon dioxide, bromine and graphite.

Property	silicon dioxide	bromine	graphite
Melting point (°C)	1610	−7	3657
Electrical conductivity	poor	poor	good
Solubility in water	insoluble	slightly soluble	insoluble

(a) What is the structure in:
    (i)   silicon dioxide?

    (1 mark)

    (ii) graphite?

    (1 mark)

    (iii) bromine?

    (1 mark)

(b) Explain why bromine has poor electrical conductivity.

    (1 mark)

(c) Explain why graphite has good electrical conductivity.

    (1 mark)

(d) Bromine is a liquid at room temperature (20 °C).
    Explain why bromine has such a low melting point compared with
    silicon dioxide and graphite.

    (2 marks)

Exam Questions

3   The diagram below shows the arrangement of atoms in pure iron.

Steel is an alloy of iron and carbon.
(a)   Draw a similar diagram to show the arrangement of atoms in steel.

(2 marks)

(b)   Steel is harder than iron. Explain why.

(3 marks)

4   The table gives data for some physical properties of a selection of substances.

Substance	Melting point	Boiling point	Electrical conductivity
A	-219	-183	poor
B	3550	4827	poor
C	1495	2870	good
D	801	1413	good when molten

(a)   What state would you expect substance D to be at room temperature?

(1 mark)

(b)   What is the structure of:
(i)   substance B?

(1 mark)

(ii)   substance D?

(1 mark)

(c)   Substance A is oxygen, O_2.
Draw a dot and cross diagram to show the outer electrons in an oxygen molecule.

(2 marks)

(d)   Substance C is a metal. According to the table, it is a good conductor of electricity.
(i)   Explain why this is.

(2 marks)

(ii)   Would substance C be a good conductor of heat?

(1 mark)

New Materials

New materials are continually being developed, with new properties. The two groups of materials covered on these pages are <u>smart materials</u> and <u>nanoparticles</u>.

Smart Materials have Some **Really Weird** Properties

1) <u>Smart</u> materials <u>behave differently</u> depending on the <u>conditions</u>, e.g. temperature.

2) A good example is <u>nitinol</u> — a "<u>shape memory alloy</u>".
   It's a metal <u>alloy</u> (about half nickel, half titanium) but when it's cool you can <u>bend it</u> and <u>twist it</u> like rubber. Bend it too far, though, and it stays bent. But here's the really clever bit — if you heat it above a certain temperature, it goes back to a "<u>remembered</u>" shape.

3) It's really handy for <u>glasses frames</u>. If you accidentally bend them, you can just pop them into a bowl of hot water and they'll <u>jump</u> back <u>into shape</u>.

4) Nitinol is also used for <u>dental braces</u>. In the mouth it <u>warms</u> and tries to return to a '<u>remembered</u>' shape, and so it gently <u>pulls the teeth</u> with it.

Nanoparticles are Really Really Really Really **Tiny**   ...smaller than that.

1) Really tiny particles, <u>1–100 nanometres</u> across, are called 'nanoparticles'
   (1 nm = 0.000 000 001 m).

2) Nanoparticles contain roughly <u>a few hundred atoms</u>.

3) Nanoparticles include <u>fullerenes</u>. These are molecules of <u>carbon</u>, shaped like <u>hollow balls</u> or <u>closed tubes</u>. The carbon atoms are arranged in <u>hexagonal rings</u>. Different fullerenes contain <u>different numbers</u> of carbon atoms.

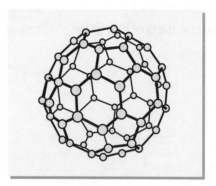

4) A nanoparticle has very <u>different properties</u> from the 'bulk' chemical that it's made from — e.g. <u>fullerenes</u> have different properties from big <u>lumps of carbon</u>.

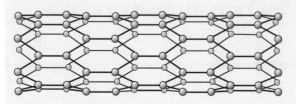

1) Fullerenes can be joined together to form <u>nanotubes</u> — teeny tiny hollow carbon tubes, a few nanometres across.

2) All those covalent bonds make carbon nanotubes <u>very strong</u>. They can be used to reinforce graphite in <u>tennis rackets</u>.

New Materials

Nanomaterials are Becoming More and More Widely Used

Using nanoparticles is known as <u>nanoscience</u>. Many <u>new uses</u> of nanoparticles are being developed.

1) They have a <u>huge surface area to volume ratio</u>, so they could help make new industrial <u>catalysts</u> (see page 158).

2) You can use nanoparticles to make <u>sensors</u> to detect one type of molecule and nothing else. These <u>highly specific</u> sensors are already being used to test water purity.

3) Nanotubes can be used to make <u>stronger</u>, <u>lighter</u> building materials.

4) New cosmetics, e.g. <u>sun tan cream</u> and <u>deodorant</u>, have been made using nanoparticles. The small particles do their job but don't leave <u>white marks</u> on the skin.

5) <u>Nanomedicine</u> is a hot topic. The idea is that tiny fullerenes are <u>absorbed</u> more easily by the body than most particles. This means they could <u>deliver drugs</u> right into the cells where they're needed.

6) New <u>lubricant coatings</u> are being developed using fullerenes. These coatings reduce friction a bit like <u>ball bearings</u> and could be used in all sorts of places from <u>artificial joints</u> to <u>gears</u>.

7) Nanotubes <u>conduct</u> electricity, so they can be used in tiny <u>electric circuits</u> for computer chips.

Bendy specs, tennis rackets and computer chips — cool...

Some nanoparticles have really <u>unexpected properties</u>. Silver's normally very unreactive, but silver nanoparticles can kill bacteria. Gold nanoparticles aren't gold-coloured — they're either red or purple. On the flipside, we also need to watch out for any unexpected harmful properties.

Polymers

Plastics are polymers. They're made up of lots of molecules joined together in long chains.

Forces Between Molecules Determine the Properties of Polymers

Strong covalent bonds hold the atoms together in long chains. But it's the bonds between the different molecule chains that determine the properties of the plastic.

Weak Forces:

Individual tangled chains of polymers, held together by weak intermolecular forces, are free to slide over each other.

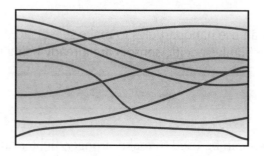

1) THERMOSOFTENING POLYMERS don't have cross-linking between chains.

2) The forces between the chains are really easy to overcome, so it's easy to melt the plastic.

3) When it cools, the polymer hardens into a new shape.

4) You can melt these plastics and remould them as many times as you like.

Strong Forces:

Some plastics have stronger intermolecular forces between the polymer chains, called crosslinks, that hold the chains firmly together.

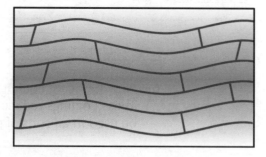

1) THERMOSETTING POLYMERS have crosslinks.

2) These hold the chains together in a solid structure.

3) The polymer doesn't soften when it's heated.

4) Thermosetting polymers are strong, hard and rigid.

Polymers

There's plastic and there's... well, plastic. You wouldn't want to make a chair with the same plastic that gets used for flimsy old carrier bags. But whatever the plastic, it's always a polymer.

How you *Make* a *Polymer* Affects its *Properties*

1) The starting materials and reaction conditions will both affect the properties of a polymer.

2) Two types of polythene can be made using different conditions:

> • Low density (LD) polythene is made by heating ethene to about 200 °C under high pressure. It's flexible and is used for bags and bottles.
>
> • High density (HD) polythene is made at a lower temperature and pressure (with a catalyst). It's more rigid and is used for water tanks and drainpipes.

The *Use* of a Plastic Depends on its *Properties*

Because plastics have different properties, you use different ones for different things. For example:

Choose from the table the plastic that would be best suited for making:

a) a disposable cup for hot drinks,

b) clothing,

c) a measuring cylinder.

Give reasons for each choice.

Plastic	Cost	Resistance to chemicals	Melting point	Transparency	Rigidity	Can be made into fibres
W	High	High	High	Low	High	No
X	Low	Low	Low	Low	Low	Yes
Y	High	High	High	High	High	No
Z	Low	Low	High	High	High	No

Answers

a) Z — low cost (disposable) and high melting point (for hot drinks),

b) X — flexible (essential for clothing) and able to be made into fibres (clothing is usually woven),

c) Y — transparent and resistant to chemicals (you need to be able to see the liquid inside and the liquid and measuring cylinder mustn't react with each other).

Polymers have different uses

Different reaction conditions result in polymers with different properties (like low density and high density polythene). If you're given information about the properties of a certain polymer and have to explain why it's suited to its use, just use your common sense — you wouldn't make a rain coat out of something that dissolves in water, and you wouldn't use a stiff, inflexible polymer to make clothes.

Warm-Up and Exam Questions

Warm-Up Questions

1) What is nitinol?
2) What does a plastic's melting point tell you about the forces between its polymer chains?
3) What are intermolecular forces between polymer chains called?
4) Give two things that can affect the properties of a polymer.

Exam Questions

1   Scientists have developed new materials using nanoparticles, which show different properties from the same materials in bulk.

   (a) Use words from the box to help you complete the sentences below.

volume	mm	catalysts	surface area	nm	circuits

   (i)   Nanoparticles are up to 100 .... in size.

   (1 mark)

   (ii) Nanoparticles have a large .... to

   .... ratio.

   (2 marks)

   (b) An elite cyclist has a bike with a frame weighing about 1 kg.
        Carbon nanotubes (CNT) were used in the manufacture of the frame of the bike.

   (i)   Suggest one property of a material made using CNTs that would make it suitable for use in a bike frame.

   (1 mark)

   (ii) Give the name of a type of molecule that can be joined together to make carbon nanotubes.

   (1 mark)

   (c) Give one other application of nanoparticles.

   (1 mark)

2   The table below shows the properties of three polymers, **A**, **B** and **C**.

   Give the polymer that would be best suited for each of the following uses:

Polymer	Properties
A	heat resistant and strong
B	very flexible and biodegradable
C	very rigid

   (a) sandwich bag

   (1 mark)

   (b) drainpipe

   (1 mark)

   (c) disposable cup

   (1 mark)

Relative Formula Mass

The biggest trouble with <u>relative atomic mass</u> and <u>relative formula mass</u> is that they <u>sound</u> so blood-curdling. Take a few deep breaths, and just enjoy, as the mists slowly clear...

Relative Atomic Mass, A_r, is Easy

1) This is just a way of saying how <u>heavy</u> different atoms are <u>compared</u> with the mass of an atom of carbon-12. So carbon-12 has A_r of <u>exactly 12</u>.
2) It turns out that the <u>relative atomic mass</u> A_r is usually just the same as the <u>mass number</u> of the element.
3) In the periodic table, the elements all have <u>two</u> numbers. The smaller one is the atomic number (how many protons it has). But the <u>bigger one</u> is the <u>mass number</u> or <u>relative atomic mass</u>.

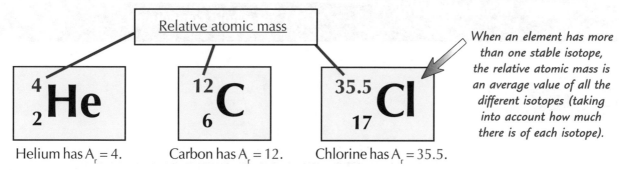

Relative atomic mass

$^{4}_{2}$He $^{12}_{6}$C $^{35.5}_{17}$Cl

When an element has more than one stable isotope, the relative atomic mass is an average value of all the different isotopes (taking into account how much there is of each isotope).

Helium has $A_r = 4$.   Carbon has $A_r = 12$.   Chlorine has $A_r = 35.5$.

Relative Formula Mass, M_r, is Also Easy

If you have a compound like $MgCl_2$ then it has a <u>relative formula mass</u>, M_r, which is just all the relative atomic masses <u>added together</u>.
For $MgCl_2$ it would be:

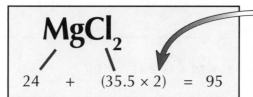

$$MgCl_2$$
$$24 + (35.5 \times 2) = 95$$

The relative atomic mass of chlorine is multiplied by 2 because there are two chlorine atoms.

So M_r for $MgCl_2$ is simply <u>95</u>.

You can easily get A_r for any element from the periodic table (see inside front cover), but in a lot of questions they give you them anyway. And that's all it is. A big fancy name like <u>relative formula mass</u> and all it means is "<u>add up all the relative atomic masses</u>".

"ONE MOLE" of a Substance is Equal to its M_r in Grams

The <u>relative formula mass</u> (A_r or M_r) of a substance <u>in grams</u> is known as <u>one mole</u> of that substance.

<u>Examples:</u>
Iron has an A_r of 56.   So one mole of iron weighs exactly 56 g
Nitrogen gas, N_2, has an M_r of 28 (2×14).   So one mole of N_2 weighs exactly 28 g

You can convert between moles and grams using this formula:

$$\text{NUMBER OF MOLES} = \frac{\text{Mass in g (of element or compound)}}{M_r \text{ (of element or compound)}}$$

<u>Example:</u> How many moles are there in 42 g of carbon?
<u>Answer:</u> No. of moles = Mass (g) / M_r = 42/12 = <u>3.5 moles</u>   Easy Peasy

Formula Mass Calculations

Although relative atomic mass and relative formula mass are <u>easy enough</u>, it can get just a tad <u>trickier</u> when you start getting into other calculations which use them. It depends on how good your maths is basically, because it's all to do with ratios and percentages.

Calculating % Mass of an Element in a Compound

This is actually dead easy — so long as you can use this formula:

$$\text{Percentage mass of an element in a compound} = \frac{A_r \times \text{no. of atoms (of that element)}}{M_r \text{ (of whole compound)}} \times 100$$

Example:

Find the percentage mass of sodium in sodium carbonate, Na_2CO_3.

<u>ANSWER:</u>

- A_r of sodium = 23
- A_r of carbon = 12
- A_r of oxygen = 16

M_r of Na_2CO_3 = $(2 \times 23) + 12 + (3 \times 16) = 106$

Now use the formula:

$$\text{Percentage Mass} = \frac{A_r \times \text{No. of atoms}}{M_r} \times 100 = \frac{23 \times 2}{106} \times 100 = 43.4\%$$

And there you have it.
Sodium makes up <u>43.4%</u> of the mass of sodium carbonate.

You can't just read these pages — work through the examples too

As usual with these calculations, <u>practice makes perfect</u>. Try these:

Find the percentage mass of oxygen in each of these:

a) Fe_2O_3
b) H_2O
c) $CaCO_3$
d) H_2SO_4

Answers on page 273.

Formula Mass Calculations

Finding the **Empirical Formula** (from Masses or Percentages)

This also sounds a lot worse than it really is. Try this for a nice simple <u>stepwise method</u>:

1) <u>List all the elements</u> in the compound (there are usually only two or three).

2) <u>Underneath them</u>, write their <u>experimental masses or percentages</u>.

3) <u>Divide</u> each mass or percentage <u>by the A_r</u> for that particular element.

4) Turn the numbers you get into <u>a nice simple ratio</u>
by multiplying and/or dividing them by well-chosen numbers.

5) Get the ratio in its <u>simplest form</u>, and that tells you the <u>empirical formula</u> of the compound.

Example:

Find the empirical formula of the iron oxide produced when
44.8 g of iron react with 19.2 g of oxygen.
(A_r for iron = 56, A_r for oxygen = 16)

<u>METHOD:</u>

1) <u>List the two elements:</u>	**Fe**	**O**
2) Write in the <u>experimental masses</u>:	44.8	19.2
3) <u>Divide by the A_r</u> for each element:	$\frac{44.8}{56} = 0.8$	$\frac{19.2}{16} = 1.2$
4) Multiply by 10...	8	12
...then divide by 4:	2	3

5) So the <u>simplest formula</u> is 2 atoms of Fe to 3 atoms of O, i.e. Fe_2O_3.

You need to realise that this <u>empirical method</u> (i.e. based on <u>experiment</u>) is the <u>only way</u>
of finding out the formula of a compound. Rust is iron oxide, sure, but is it FeO, or Fe_2O_3?
Only an experiment to determine the empirical formula will tell you for certain.

It's all about making the ratio as simple as possible...

If you find these scary, just keep practising using the stepwise method until you've mastered it.
Try this:
Find the empirical formula of the compound formed from 2.4 g of carbon and 0.8 g of hydrogen*.

Calculating Masses in Reactions

You can also work out masses of reactants (starting materials) and products in reactions.

The Three Important Steps — *Not to be Missed...*

1) Write out the balanced equation.

2) Work out the M_r — just for the two bits you want.

3) Apply the rule: Divide to get one, then multiply to get all.
(But you have to apply this first to the substance they
give you information about, and then the other one.)

Don't worry — these steps should all make sense when you look at the example below.

Example:

What mass of magnesium oxide is produced when 60 g of magnesium is burned in air?

Answer:

1) Write out the balanced equation: $2Mg + O_2 \rightarrow 2MgO$

2) Work out the relative formula masses: $2 \times 24 \rightarrow 2 \times (24 + 16)$
(don't do the oxygen — you don't need it) $48 \rightarrow 80$

3) Apply the rule: Divide to get one, then multiply to get all:

**The two numbers, 48 and 80, tell us that 48 g of Mg react to give 80 g of MgO.
Here's the tricky bit. You've now got to be able to write this down:**

> 48 g of Mg ...... reacts to give ...... 80g of MgO
> 1 g of Mg ...... reacts to give ......
> 60 g of Mg ...... reacts to give ......

The big clue is that in the question they've said we want to burn "60 g of magnesium",
i.e. they've told us how much magnesium to have, and that's how you know to write down
the left-hand side of it first, because:

**We'll first need to ÷ by 48 to get 1 g of Mg
and then need to × by 60 to get 60 g of Mg.**

Then you can work out the numbers on the other side (shown in blue below) by realising
that you must divide both sides by 48 and then multiply both sides by 60.

÷ 48 48 g of Mg .... 80 g of MgO ÷ 48
1 g of Mg .... 1.67 g of MgO
× 60 60 g of Mg .... 100 g of MgO × 60

The mass of product is called the yield of a reaction. You should realise that in practice you never get 100% of the yield, so the amount of product will be slightly less than calculated (see page 133).

This finally tells us that 60 g of magnesium will produce 100 g of magnesium oxide.

If the question had said "Find how much magnesium gives 500 g of magnesium oxide",
you'd fill in the MgO side first, because that's the one you'd have the information about.

Warm-Up and Exam Questions

Lots to remember on those four pages. Try these and see how good your understanding really is.

Warm-Up Questions

1) What name is given to how heavy different atoms are compared with the mass of an atom of carbon-12?
2) What name is given to the sum of the relative atomic masses of the atoms in a molecule?
3) Write down the definition of a mole.
4) What is the mass of one mole of oxygen gas?

Exam Questions

1 (a) Boron has two main isotopes, $^{11}_{5}B$ and $^{10}_{5}B$. Its A_r value is 10.8.
   (i)   What does A_r stand for?

(1 mark)

   (ii) What is the difference in structure between the two boron isotopes?

(1 mark)

   (iii) Explain why the A_r isn't a whole number.

(1 mark)

   (b) Use the A_r values B = 11, O = 16, F = 19 and H = 1 to calculate the relative formula masses of these boron compounds:
   (i) BF_3

(1 mark)

   (ii) $B(OH)_3$.

(1 mark)

2 Analysis of an oxide of sulfur shows that it contains 60% oxygen by mass.
   (A_r values: S = 32, O = 16.)
   (a) What is the percentage mass of sulfur in the oxide?

(1 mark)

   (b) Work out the formula of the oxide.

(2 marks)

3 Heating a test tube containing 2 g of calcium carbonate produced 1.08 g of calcium oxide when it was reweighed. The equation for the reaction is:
   $$CaCO_3(s) \rightarrow CaO(s) + CO_2(g)$$
   (M_r values: $CaCO_3$ = 100, CaO = 56.)

   (a) Calculate the amount of calcium oxide you would expect to be formed from 2 g of calcium carbonate.

(1 mark)

   (b) Compare the value to the mass obtained in the experiment.
        Suggest a possible reason for the difference.

(1 mark)

Atom Economy

It's important in industrial reactions that as much of the reactants as possible get turned into useful products. This depends on the <u>atom economy</u> and the <u>percentage yield</u> (see page 132) of the reaction.

"Atom Economy" — % of Reactants Changed to Useful Products

1) A lot of reactions make <u>more than one product</u>. Some of them will be <u>useful</u>, but others will just be <u>waste</u>, e.g. when you make quicklime from limestone, you also get CO_2 as a waste product.

2) The <u>atom economy</u> of a reaction tells you how much of the <u>mass</u> of the reactants is wasted when manufacturing a chemical.

Here's the equation:

$$\text{atom economy} = \frac{\text{total } M_r \text{ of desired products}}{\text{total } M_r \text{ of all products}} \times 100$$

3) <u>100%</u> atom economy means that <u>all</u> the atoms in the reactants have been turned into <u>useful</u> (desired) <u>products</u>. The <u>higher</u> the atom economy the '<u>greener</u>' the process.

Example

Hydrogen gas is made on a large scale by reacting natural gas (methane) with steam.

$$CH_4(g) + H_2O(g) \rightarrow CO(g) + 3H_2(g)$$

Calculate the atom economy of this reaction.

<u>Method:</u>

1) <u>Identify</u> the useful product — that's the <u>hydrogen gas</u>.

2) Work out the M_r of <u>all the products</u> and the <u>useful product</u>:

CO	$3H_2$	$3H_2$
12 + 16	3 × (2 × 1)	3 × (2 × 1)
34		6

3) Use the <u>formula</u> to calculate the atom economy:

$$\text{atom economy} = \frac{6}{34} \times 100 = \underline{17.6\%}$$

So in this reaction, <u>over 80%</u> of the starting materials are <u>wasted</u>.

Atom Economy

So, a <u>high</u> atom economy is <u>good</u>, and a <u>low</u> atom economy is <u>not so good</u>.
Make sure you know all the reasons why.

High Atom Economy is Better for **Profits** and the **Environment**

1) Pretty obviously, if you're making <u>lots of waste</u>, that's a <u>problem</u>.

2) Reactions with low atom economy <u>use up resources</u> very quickly.

3) At the same time, they produce loads of <u>waste</u> materials that have to be <u>disposed</u> of somehow.

4) That tends to make these reactions <u>unsustainable</u> — the raw materials will run out and the waste has to go somewhere.

5) For the same reasons, low atom economy reactions aren't usually <u>profitable</u>.

6) Raw materials are <u>expensive to buy</u>, and waste products can be expensive to <u>remove</u> and dispose of <u>responsibly</u>.

7) The best way around the problem is to find a <u>use</u> for the waste products rather than just <u>throwing them away</u>.

8) There's often <u>more than one way</u> to make the product you want, so the trick is to come up with a reaction that gives <u>useful</u> "<u>by-products</u>" rather than useless ones.

9) The reactions with the <u>highest</u> atom economy are the ones that only have <u>one product</u>.

10) Those reactions have an atom economy of <u>100%</u>.

Atom economy — important, but not the whole story...

Atom economy isn't the only thing that affects profits — there are other costs besides buying raw materials and disposing of waste. There are <u>energy</u> and <u>equipment</u> costs, as well as the cost of <u>paying people</u> to work at the plant. You need to think about the <u>percentage yield</u> of the reaction too (p.132).

Percentage Yield

Percentage yield tells you about the <u>overall success</u> of an experiment. It compares what you calculate you should get (<u>predicted yield</u>) with what you get in practice (<u>actual yield</u>).

Percentage Yield Compares **Actual** and **Predicted** *Yield*

The amount of product you get is known as the <u>yield</u>. The more reactants you start with, the higher the <u>actual yield</u> will be — that's pretty obvious. But the <u>percentage yield doesn't</u> depend on the amount of reactants you started with — it's a <u>percentage</u>.

1) The <u>predicted yield</u> of a reaction can be calculated from the <u>balanced reaction equation</u>.

The predicted yield is sometimes called the theoretical yield.

2) Percentage yield is given by the formula:

$$\text{percentage yield} = \frac{\text{actual yield (grams)}}{\text{predicted yield (grams)}} \times 100$$

3) Percentage yield is <u>always</u> somewhere between 0 and 100%.

4) A 100% percentage yield means that you got <u>all</u> the product you expected to get.

5) A 0% yield means that <u>no</u> reactants were converted into product, i.e. no product at all was <u>made</u>.

Yields are Always **Less Than 100%**

Even though <u>no atoms are gained or lost</u> in reactions, in real life, you <u>never</u> get a 100% percentage yield. Some product or reactant <u>always</u> gets lost along the way — and that goes for big <u>industrial processes</u> as well as school lab experiments.

Lots of things can go wrong, but there are three examples conveniently located on the next page.

Even with the best equipment, you can't get the maximum product

It's always quite sad when you get to the end of a really long lab experiment and it turns out you've lost half of your product in the filter paper — it brings a tear to my eye just thinking about it. It makes you feel a bit better that the same thing happens to the big boys in industry too though... C'est la vie.

Percentage Yield and Sustainable Development

Here are **Three Reasons** Why Your Yield **Won't** be 100%:

1) The Reaction is **Reversible**

A <u>reversible reaction</u> is one where the <u>products</u> of the reaction can <u>themselves react</u> to produce the <u>original reactants</u>

$$A + B \rightleftharpoons C + D$$

<u>For example:</u>
ammonium chloride $\rightleftharpoons$ ammonia + hydrogen chloride

1) This means that the reactants will never be completely converted to products because the reaction goes both ways.
2) Some of the <u>products</u> are always <u>reacting together</u> to change back to the original reactants.
3) This will mean a <u>lower yield</u>.

2) **Filtration**

1) When you <u>filter a liquid</u> to remove <u>solid particles</u>, you nearly always <u>lose</u> a bit of liquid or a bit of solid.
2) So, some of the product may be lost when it's <u>separated</u> from the reaction mixture.

3) **Unexpected Reactions**

1) Things don't always go exactly to plan. Sometimes there can be other <u>unexpected reactions</u> happening which <u>use up the reactants</u>.
2) This means there's not as much reactant to make the <u>product</u> you want.

Product Yield is Important for **Sustainable Development**

1) Thinking about product yield is important for <u>sustainable development</u>.

2) Sustainable development is about making sure that we don't use <u>resources</u> faster than they can be <u>replaced</u> — there needs to be enough for <u>future generations</u> too.

3) So, for example, using as <u>little energy</u> as possible to create the <u>highest product yield possible</u> means that resources are <u>saved</u>. A low yield means wasted chemicals — not very sustainable.

Warm-Up and Exam Questions

Time to see what you can remember about atom economy and percentage yield...

Warm-Up Questions

1) What effect does a waste by-product have on the atom economy of a reaction?
2) What is the atom economy of the reaction shown? $2SO_2 + O_2 \rightarrow 2SO_3$
3) Why might a reaction with a low atom economy be bad for the environment?
4) What is the percentage yield of a reaction which produced 4 g of product if the predicted yield was 5 g?
5) Why might a reaction with a low percentage yield be bad for sustainable development?

Exam Questions

1   Ethanol produced by the fermentation of sugar can be converted into ethene, as shown below. The ethene can then be used to make polythene.

C_2H_6O (g) $\rightarrow$ C_2H_4 (g) + H_2O (g)

Calculate the atom economy of this reaction. (A_r values: C = 12, O = 16, H = 1.)

(3 marks)

2   A sample of copper was made by reducing 4 g of copper oxide with methane gas. When the black copper oxide turned orange-red, the sample was scraped out into a beaker. Sulfuric acid was added to dissolve any copper oxide that remained. The sample was then washed, filtered and dried. 2.8 g of copper was obtained. (A_r values: Cu = 63.5, O = 16.)

The equation for this reaction is: $CH_4 + 4CuO \rightarrow 4Cu + 2H_2O + CO_2$

(a)   Use the equation to calculate the maximum mass of copper which could be obtained from the reaction (the predicted yield).

(3 marks)

(b)   Calculate the percentage yield of the reaction.

(2 marks)

3   *In this question you will be assessed on the quality of your English, the organisation of your ideas and your use of appropriate specialist vocabulary.*

Discuss the reasons why yields from chemical reactions are always less than 100%.

(6 marks)

4   Which of the following statements about atom economy is **not** true?

A   Reactions that only have one product have a very high atom economy.
B   Reactions with a low atom economy use up resources quickly.
C   Reactions with a low atom economy are not usually profitable.
D   Reactions with a high atom economy use up resources quickly.

(1 mark)

Revision Summary for Section 4

Some people skip these pages. But what's the point in reading that great big section if you're not going to check if you really know it or not? Look, just read the first ten questions, and I guarantee there'll be an answer you'll have to look up. And when it comes up in the exam, you'll be so glad you did.

1) What do the mass number and atomic number represent?

2) Draw a table showing the relative masses of the three types of particle in an atom.

3) What is a compound?

4) What size groups did Döbereiner organise the elements into?

5) Give two reasons why Newlands' Octaves were criticised.

6) Why did Mendeleev leave gaps in his Table of Elements?

7) What feature of atoms determines the order of the periodic table?

8) What are groups in the periodic table? Explain their significance in terms of electrons.

9) Which group contains the alkali metals? How many electrons do they each have in their outer shell?

10) Give details of the reactions of the alkali metals with water.

11)*Write a balanced equation for the reaction between lithium and chlorine.

12) Name a type of experiment that could be used to determine the relative reactivity of the halogens.

13) Describe the process of ionic bonding.

14) Describe the structure of a crystal of sodium chloride.

15) List the main properties of ionic compounds.

16)*Use information from the periodic table to help you work out the formula of calcium chloride.

17)* Draw a diagram to show the electronic structure of an Mg^{2+} ion (magnesium's atomic number is 12).

18) Sketch dot and cross diagrams showing the bonding in molecules of:
    a) hydrogen,   b) hydrogen chloride,   c) water,   d) ammonia

19) What are the two types of covalent substance? Give three examples of each.

20) List three properties of metals and explain how metallic bonding causes these properties.

21) Give an example of a "smart" material and describe how it behaves.

22) What are nanoparticles?

23) Explain the difference between thermosoftening and thermosetting polymers.

24) Define relative atomic mass and relative formula mass.

25)*Find A_r or M_r for these (use the periodic table at the front of the book):
    a) Ca   b) Ag   c) CO_2 d) $MgCO_3$ e) Na_2CO_3 f) ZnO   g) KOH   h) NH_3

26) What is the link between moles and relative formula mass?

27)*a) Calculate the percentage mass of carbon in: i) $CaCO_3$ ii) CO_2 iii) CH_4
    b) Calculate the percentage mass of metal in: i) Na_2O ii) Fe_2O_3 iii) Al_2O_3

28)*What is an empirical formula? Find the empirical formula of the compound formed when 21.9 g of magnesium, 29.2 g of sulfur and 58.4 g of oxygen react.

29)*What mass of sodium is needed to produce 108.2 g of sodium oxide (Na_2O)?

30) Write the equation for calculating the atom economy of a reaction.

31) Explain why it is important to use industrial reactions with a high atom economy.

* Answers on page 274.

Identifying Positive Ions

Say you've got a compound, but you don't know what it is. Well, you'd want to identify it...

Flame Tests — Spot the Colour

Compounds of some <u>metals</u> give a characteristic <u>colour</u> when heated.
This is the idea behind <u>flame tests</u>.

> <u>Sodium</u>, Na^+, gives an orange/yellow flame.
> <u>Potassium</u>, K^+, gives a lilac flame.
> <u>Calcium</u>, Ca^{2+}, gives a brick-red flame.
> <u>Copper</u>, Cu^{2+}, gives a blue-green flame.

You can use these <u>colours</u> to <u>detect</u> and <u>identify</u> different ions.

Add Sodium Hydroxide and Look for a Coloured Precipitate

1) A precipitation reaction is where two solutions react to form
   an <u>insoluble solid compound</u> called a <u>precipitate</u>.

2) Many <u>metal hydroxides</u> are <u>insoluble</u> and precipitate out of solution when you add an alkali.

3) Some of these hydroxides have a <u>characteristic colour</u>.

4) So in this test you add a few drops of <u>sodium hydroxide</u> solution to a solution
   of your mystery compound — in the hope of forming an insoluble hydroxide.

5) If you get a <u>coloured insoluble hydroxide</u> you can then identify the metal ion that was in the compound.

"Metal"	Colour of precipitate	Ionic Reaction
Calcium, Ca^{2+}	White	$Ca^{2+}_{(aq)} + 2OH^-_{(aq)} \rightarrow Ca(OH)_{2(s)}$
Copper(II), Cu^{2+}	Blue	$Cu^{2+}_{(aq)} + 2OH^-_{(aq)} \rightarrow Cu(OH)_{2(s)}$
Iron(II), Fe^{2+}	Sludgy green	$Fe^{2+}_{(aq)} + 2OH^-_{(aq)} \rightarrow Fe(OH)_{2(s)}$
Iron(III), Fe^{3+}	Reddish brown	$Fe^{3+}_{(aq)} + 3OH^-_{(aq)} \rightarrow Fe(OH)_{3(s)}$
Zinc, Zn^{2+}	White at first. But then redissolves in excess NaOH to form a colourless solution.	$Zn^{2+}_{(aq)} + 2OH^-_{(aq)} \rightarrow Zn(OH)_{2(s)}$ then $Zn(OH)_{2(s)} + 2OH^-_{(aq)} \rightarrow Zn(OH)_4^{2-}_{(aq)}$

*NaOH isn't the only solution that can be used for precipitation
reactions — you can use other solutions of ionic compounds too.*

Ionic Equations Show Just the Useful Bits of Reactions

1) The reactions in the table above are <u>ionic equations</u>.
   Ionic equations are 'half' a full equation, if you like. For example:

$$Ca^{2+}_{(aq)} + 2OH^-_{(aq)} \longrightarrow Ca(OH)_{2(s)}$$

2) This <u>just</u> shows the formation of (solid) <u>calcium hydroxide</u> from the <u>calcium ions</u> and the
   <u>hydroxide ions</u> in solution. And it's the formation of this that helps identify the compound.
   The <u>full</u> equation in the above reaction would be (if you started off with <u>calcium chloride</u>, say):

$$CaCl_{2(aq)} + 2NaOH_{(aq)} \longrightarrow Ca(OH)_{2(s)} + 2NaCl_{(aq)}$$

3) But the formation of <u>sodium chloride</u> is of no great interest here
   — it's not helping <u>identify</u> the compound.

4) So the ionic equation just concentrates on the <u>good bits</u>.

Identifying Negative Ions

So now maybe you know what the <u>positive</u> part of your mystery substance is (see previous page). Now it's time to test for the <u>negative</u> bit.

Hydrochloric Acid Can Help Detect *Carbonates*

With dilute <u>hydrochloric acid</u>, <u>carbonates</u> (CO_3^{2-}) will fizz because they give off <u>carbon dioxide</u>.

$$CO_3^{2-}(s) + 2H^+(aq) \longrightarrow CO_2(g) + H_2O(l)$$

You can test for carbon dioxide using <u>limewater</u>.

The *Limewater* Test

Carbon dioxide <u>turns limewater cloudy</u> — just bubble the gas through a test tube of limewater and watch what happens. If the water goes cloudy you've identified a <u>carbonate ion</u>.

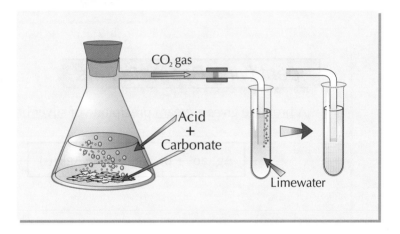

Test for *Sulfates* with HCl and *Barium Chloride*

To identify a <u>sulfate</u> ion (SO_4^{2-}), add dilute HCl, followed by <u>barium chloride solution</u>, $BaCl_2$(aq).

$$Ba^{2+}(aq) + SO_4^{2-}(aq) \longrightarrow BaSO_4(s)$$

A <u>white precipitate</u> of <u>barium sulfate</u> means the original compound was a sulfate.

(The <u>hydrochloric acid</u> is added to get rid of any traces of <u>carbonate</u> ions before you do the test. These would also produce a precipitate, so they'd <u>confuse</u> the results.)

Detect sulfate ions with hydrochloric acid and barium chloride

Being able to identify unknown substances has useful applications in the <u>real world</u>...
For example, tests like these could help to identify substances at a crime scene. However, you're most likely to come across them in the lab, being used by frighteningly enthusiastic chemistry teachers.

Identifying Negative Ions

There's one more test for negative ions on this page. Enjoy...

Test for *Halides (Cl⁻, Br⁻, I⁻)* with Nitric Acid and *Silver Nitrate*

To identify a <u>halide</u> ion, add dilute <u>nitric acid</u> (HNO_3), followed by <u>silver nitrate solution</u>, $AgNO_3$(aq).

Identifying **Chloride** Ions

A <u>chloride</u> gives a <u>white</u> precipitate of <u>silver chloride</u>.

$$Ag^+(aq) + Cl^-(aq) \longrightarrow AgCl(s)$$

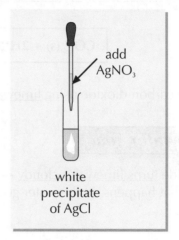

add
AgNO₃

white
precipitate
of AgCl

Identifying **Bromide** Ions

A <u>bromide</u> gives a <u>cream</u> precipitate of <u>silver bromide</u>.

$$Ag^+(aq) + Br^-(aq) \longrightarrow AgBr(s)$$

cream
precipitate
of AgBr

Identifying **Iodide** Ions

An <u>iodide</u> gives a <u>yellow</u> precipitate of <u>silver iodide</u>.

$$Ag^+(aq) + I^-(aq) \longrightarrow AgI(s)$$

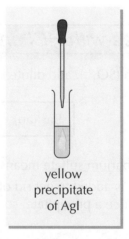

yellow
precipitate
of AgI

(Again, the <u>acid</u> is added to get rid of <u>carbonate</u> ions before the test.
You use <u>nitric acid</u> in this test, though, <u>not HCl</u>.)

It may need loads of tests to work out what a substance is...

If you're given the <u>results</u> from several chemical tests, and you have to say what the substance is, make sure you write down the thinking behind your answer — you never know, it could get you an extra mark.

Paper Chromatography

Now time for some proper chemistry. You can sometimes tell what substances are present in a mixture using <u>chromatography</u>.

Artificial Colours *Can be Separated Using Paper* Chromatography

A <u>food colouring</u> might contain <u>one dye</u> or it might be a <u>mixture of dyes</u>.

Here's how you can tell:

1) <u>Extract</u> the colour from a food sample by placing it in a small cup with a few drops of <u>solvent</u> (can be water, ethanol, salt water, etc).

2) Put <u>spots</u> of the coloured solution on a <u>pencil baseline</u> on filter paper.
(Don't use pen because it might dissolve in the solvent and confuse everything.)

3) Roll up the sheet and put it in a <u>beaker</u> with some <u>solvent</u>
— but keep the baseline above the level of the solvent.

4) The solvent <u>seeps</u> up the paper, taking the dyes with it.
Different dyes form spots in <u>different places</u>.

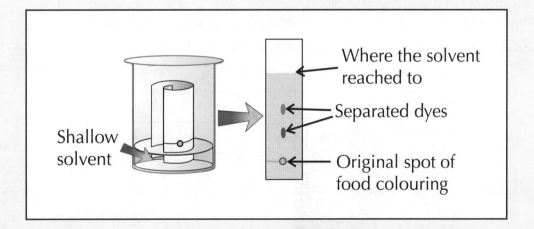

5) Watch out though — a chromatogram with <u>four spots</u> means <u>at least four</u> dyes, not exactly four dyes.
There <u>could</u> be <u>five</u> dyes, with two of them making a spot in the same place.
It <u>can't be three</u> dyes though, because one dye can't split into two spots.

Make sure your solvent level is lower than the spots...

Chromatography is really useful — and not just for separating food colourings. It can separate accurately even dead complex mixtures if you choose the right solvent, paper and conditions.

Chemical Analysis and Instrumental Methods

Nowadays there are some pretty clever ways of identifying substances. We don't have to stick to filter paper — we can use machines as well.

Machines Can Also Analyse Unknown Substances

You can identify elements and compounds using instrumental methods — this just means using machines.

> Advantages of Using Machines
> 1) Very sensitive — can detect even the tiniest amounts of substances.
> 2) Very fast and tests can be automated.
> 3) Very accurate.

Gas Chromatography Can be Used to Identify Substances

Gas chromatography can separate out a mixture of compounds and help you identify the substances present.

1) A gas is used to carry substances through a column packed with a solid material.

2) The substances travel through the tube at different speeds, so they're separated.

3) The time they take to reach the detector is called the retention time. It can be used to help identify the substances.

4) The recorder draws a gas chromatograph. The number of peaks shows the number of different compounds in the sample.

5) The position of the peaks shows the retention time of each substance.

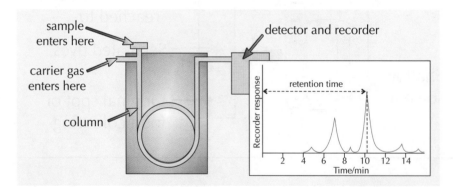

6) The gas chromatography column can also be linked to a mass spectrometer. This process is known as GC-MS and can identify the substances leaving the column very accurately.

7) You can work out the relative molecular mass of each of the substances from the graph it draws. You just read off from the molecular ion peak.

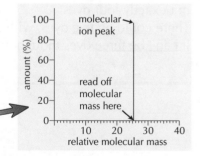

Unfortunately, machines can't do the exam for you...

Make sure you don't get the two types of chromatography muddled up... there's paper and then there's gas. Gas chromatography can tell you a lot more information about substances.

Warm-Up and Exam Questions

Lots of chemical detective work over the last five pages. Try these questions to test your memory.

Warm-Up Questions

1) What colour is the precipitate formed when NaOH is added to a solution of copper(II) ions?
2) How would you test for carbon dioxide?
3) Give three advantages of using instrumental methods to identify elements and compounds.
4) How can gas chromatography be used to work out the relative molecular mass of a substance?

Exam Questions

1   Kelly carried out flame tests on compounds of four different metal ions. Complete the table on the right showing her results.

Flame colour	Metal ion
green	
	K^+
yellow	
	Ca^{2+}

(4 marks)

2   The table below shows the results of a series of chemical tests conducted on two unknown compounds, X and Y.

TEST	OBSERVATION	
	COMPOUND X	COMPOUND Y
sodium hydroxide solution	white precipitate	no precipitate
hydrochloric acid & barium chloride solution	no precipitate	no precipitate
flame test	brick-red flame	lilac flame
nitric acid & silver nitrate solution	white precipitate	yellow precipitate

(a)   What is the chemical name of compound X?

(2 marks)

(b)   What is the chemical name of compound Y?

(2 marks)

3   Scientists analysed the composition of six food colourings using chromatography. Four of the colourings were unknown (**1 – 4**), and the other two were known, sunrise yellow and sunset red. The results are shown below.

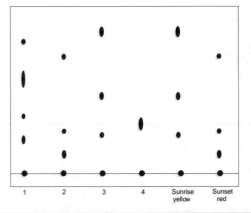

(a)   Which food colouring is most likely to be a pure compound?

(1 mark)

(b)   Which food colouring contains at least four different compounds?

(1 mark)

(c)   Which food colouring contains the same compounds as sunrise yellow?

(1 mark)

Industrial Chemical Synthesis

Chemical synthesis is the term used by those in the know to describe the process of making complex chemical compounds from simpler ones. The chemical industry does this every day...

The Chemical Industry Makes **Useful** Products

Chemicals aren't just found in the laboratory. Most of the products you come across in your day-to-day life will have been carefully researched, formulated and tested by chemists. Here are a few examples...

Food additives — the chemical industry produces additives like preservatives, colourings and flavourings for food producers.

Cleaning and decorating products — things like paints contain loads of different pigments and dyes, all of which have been developed by chemists. Cleaning products like bleach, oven cleaner and washing-up liquid will all have been developed by chemists.

Drugs — the pharmaceutical industry is huge (see next page). Whenever you have a headache or an upset tummy the drugs you take will have gone through loads of development and testing before you get to use them.

Fertilisers — we use about a million tonnes of fertiliser every year. Amongst other things fertilisers contain loads of ammonia, all of which has to be produced by the chemical industry.

As well as figuring out how to make chemicals, chemists must also figure out how to make them in the way that produces the highest yield (p.132)). They must also think about the environment, choosing processes with a low impact.

The **Type** of **Manufacturing Process** Depends on the **Product**

Batch Production Only Operates at **Certain Times**

Pharmaceutical drugs are complicated to make and there's fairly low demand for them. Batch production is often the most cost-effective way to produce small quantities of different drugs to order, because:

1) It's flexible — several different products can be made using the same equipment.
2) Start-up costs are relatively low — small-scale, multi-purpose equipment can be bought off the shelf.

But batch production does have disadvantages:

3) It's labour-intensive — the equipment needs to be set up and manually controlled for each batch and then cleaned out at the end.
4) It can be tricky to keep the same quality from batch to batch.

Continuous Production Runs **All the Time**

Large-scale industrial manufacture of popular chemicals, e.g. the Haber process for making ammonia uses continuous production because:

1) Production never stops, so you don't waste time emptying the reactor and setting it up again.
2) It runs automatically — you only need to interfere if something goes wrong.
3) The quality of the product is very consistent.

But, start-up costs to build the plant are huge, and it isn't cost-effective to run at less than full capacity.

Industrial Chemical Synthesis

The chemical <u>industry</u> is <u>big business</u>...

The Chemical Industry is **Huge**

It's absolutely massive, in terms of both the amount of chemicals it produces
and the money it generates.

Scale — chemicals can be produced on a **large** or **small** scale

1) Some chemicals are produced on a <u>massive scale</u> — for example, over 150 million tonnes
   of <u>sulfuric acid</u> are produced around the world every year.
2) Sulfuric acid has loads of <u>different uses</u>, for example in car batteries and fertiliser production.
3) Other chemicals, e.g. <u>pharmaceuticals</u>, are produced on a <u>smaller scale</u>,
   but this <u>doesn't</u> make them any <u>less important</u> — we just need less of them.

Sectors — there are loads of different sectors within the chemical industry

1) In the UK, the chemical industry makes up a significant chunk of the <u>economy</u>.
2) In the UK alone, there are over <u>200 000 people</u> employed in the chemical industry.
3) Some chemicals are sold <u>directly</u> to <u>consumers</u>, while others are sold to other <u>industries</u>
   as raw materials for other products.
4) The <u>pharmaceutical</u> sector has the largest share of the industry.

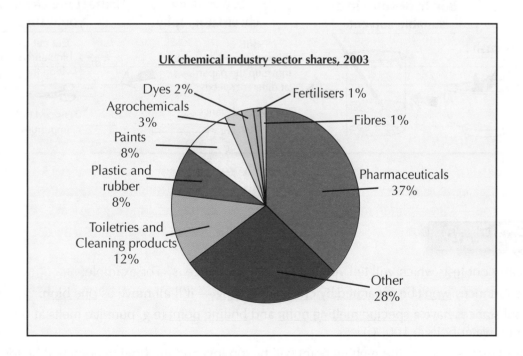

UK chemical industry sector shares, 2003

The UK chemical industry produces tonnes of pharmaceuticals

The chemical industry has certainly got its fingers in lots of different pies. It's pretty exhausting to
think about really — all those chemists locked away producing the stuff that we need every day.
But don't fall asleep just yet, there's still <u>more</u> on chemical synthesis on the next page...

Industrial Chemical Synthesis

When you're manufacturing drugs there are lots of things you have to think about...

Pharmaceutical Drugs Often Cost A Lot — For Several Reasons

Research and Development — finding a suitable compound, testing it, modifying it, testing again, until it's ready. This involves the work of lots of highly paid scientists.

Trialling — no drug can be sold until it's gone through loads of time-consuming tests including animal trials and human trials. The manufacturer has to prove that the drug meets legal requirements so it works and it's safe.

Manufacture — multi-step batch production is labour-intensive and can't be automated. Other costs include energy and raw materials. The raw materials for pharmaceuticals are often rare and sometimes need to be extracted from plants (an expensive process).

It takes about 12 years and £900 million to develop a new drug and get it onto the market. Ouch.

How to **Extract** Raw Materials from **Plants**

To extract a substance from a plant, it has to be crushed then boiled and dissolved in a suitable solvent. Then, you can extract the substance you want by chromatography.

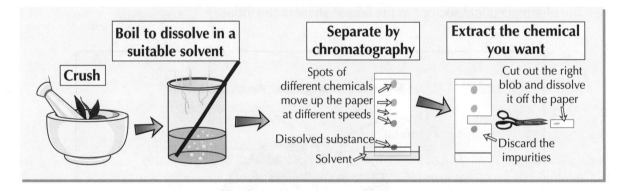

Crush → **Boil to dissolve in a suitable solvent** → **Separate by chromatography** → **Extract the chemical you want**

Spots of different chemicals move up the paper at different speeds

Dissolved substance

Solvent

Cut out the right blob and dissolve it off the paper

Discard the impurities

Test For Purity

You can carry out tests which will tell you how pure a substance is. For example:

1) Pure substances won't be separated by chromatography — it'll all move as one blob.
2) Pure substances have a specific melting point and boiling point (e.g. pure ice melts at 0 °C, and pure water boils at 100 °C).
3) If a substance is impure, the melting point will be too low and the boiling point will be too high (so if some ice melts at –2 °C, it's probably got an impurity in it e.g. salt).

I wish they'd find a drug to cure exams...

£900 million. Blimey, I wish I had that much money to throw around. But it's all worth it in the end — without pharmaceutical companies there'd be no painkillers, thousands more people would die every year and the probability of you dying if you popped across the ocean on holiday would sky-rocket.

Warm-Up and Exam Questions

Well, I'm afraid that's all you're getting about the chemical industry. Why not try these questions out to see how much you've learnt. It'd be a crying shame not to.

Warm-Up Questions

1) Give one example of a household cleaning product that will have been developed by chemists.
2) Name the type of manufacturing process that you would use for producing small quantities of drugs to order.
3) Explain why drugs need to be trialled before they are sold.
4) Give two reasons why pharmaceutical drugs often cost a lot to produce.

Exam Questions

1   A chemical company uses continuous production to produce chemicals on a large scale.
   (a)   Give an example of a chemical that is produced on a large scale.

   (1 mark)

   (b)   Give one disadvantage of producing chemicals by continuous production.

   (1 mark)

   (c)   Give two advantages of producing chemicals by continuous production.

   (2 marks)

2   *In this question you will be assessed on the quality of your English,
   the organisation of your ideas and your use of appropriate specialist vocabulary.*

   Many pharmaceutical drugs are produced from raw materials taken from plants.
   The first step in this process is to extract the raw material from the plant.
   Describe how you would extract a substance from a plant in the laboratory.

   (6 marks)

3   Sammy is trying to produce a pure sample of Chemical **Y** which has a melting point of 140 °C.
   She attempts to purify Chemical **Y** in three different ways and then tests the purity
   of the three samples produced. Her results are shown in the table below.

Sample	Number of spots produced by chromatography	melting point (°C)
1	1	139
2	3	140
3	1	140

   Which of the samples is the most pure? Explain your answer.

   (2 marks)

Rate of Reaction

Reactions can be <u>fast</u> or <u>slow</u> — you've probably already realised that. This page is about what affects the <u>rate of a reaction</u>, and the next two pages tell you what you can do to <u>measure it</u>.

Reactions Can Go At All Sorts of **Different Rates**

1) One of the <u>slowest</u> is the <u>rusting</u> of iron.

2) A <u>moderate speed</u> reaction is a <u>metal</u> (like magnesium) reacting with <u>acid</u> to produce a gentle stream of <u>bubbles</u>.

3) A <u>really fast</u> reaction is an <u>explosion</u>, where it's all over in a <u>fraction</u> of a second.

The **Rate of a Reaction** Depends on **Four Things**:

| 1) Temperature | 2) Concentration — (or <u>pressure</u> for gases) |
| 3) Catalyst | 4) Surface area of solids — (or <u>size</u> of solid pieces) |

Typical Graphs for Rate of Reaction

The plot below shows how the rate of a particular reaction varies under <u>different conditions</u>. The <u>quickest reaction</u> is shown by the line with the <u>steepest slope</u>. Also, the faster a reaction goes, the sooner it finishes, which means that the line becomes <u>flat</u> earlier.

You could also show the amount of reactant used up over time instead — the graphs would have the same shape.

1) <u>Graph 1</u> represents the original <u>fairly slow</u> reaction. The graph is not too steep.

2) <u>Graphs 2 and 3</u> represent the reaction taking place <u>quicker</u> but with the <u>same initial amounts</u>. The slope of the graphs gets steeper.

3) The <u>increased rate</u> could be due to <u>any</u> of these:

> a) increase in <u>temperature</u>
> b) increase in <u>concentration</u> (or pressure)
> c) <u>catalyst</u> added
> d) solid reactant crushed up into <u>smaller bits</u>.

Amount of product evolved

④ faster, and more reactants

end of reaction

③ much faster reaction

② faster reaction

① original reaction

Time

4) <u>Graph 4</u> produces <u>more product</u> as well as going <u>faster</u>. This can <u>only</u> happen if <u>more reactant(s)</u> are added at the start. <u>Graphs 1, 2 and 3</u> all converge at the same level, showing that they all produce the same amount of product, although they take <u>different</u> times to get there.

Controlling the **Rate of Reaction** is Important in **Industry**

In <u>industry</u>, it's important to <u>control</u> the rate of a chemical reaction for <u>two</u> main reasons:

1) <u>Safety</u> — if the reaction is <u>too fast</u> it could cause an <u>explosion</u>, which can be a bit <u>dangerous</u>.

2) <u>Economic reasons</u> — changing the conditions can be <u>costly</u>. For example, using very high <u>temperatures</u> means there'll be bigger <u>fuel bills</u>, so the cost of production is pushed up. But, a <u>faster rate</u> means that <u>more product</u> will be produced in <u>less time</u>. Companies often have to choose optimum conditions that give <u>low production costs</u>, but this may mean compromising on the <u>rate of production</u>, or the <u>yield</u>.

Measuring Rates of Reaction

If you want to know the rate of reaction then it's fairly easy to <u>measure</u> it.
There are <u>three</u> ways of measuring rate of reaction for you to look at.

Ways to **Measure the Rate** of a Reaction

The <u>rate of a reaction</u> can be observed <u>either</u> by measuring how quickly the reactants are used up
or how quickly the products are formed. It's usually a lot easier to measure <u>products forming</u>.

The rate of reaction can be calculated using the following formula:

$$\text{Rate of reaction} = \frac{\text{amount of reactant used or amount of product formed}}{\text{time}}$$

There are different ways that the rate of a reaction can be <u>measured</u>. Here are three:

1) Precipitation

1) This is when the product of the reaction is a <u>precipitate</u> which <u>clouds</u> the solution.

2) Observe a <u>mark</u> through the solution and measure how long it takes for it to <u>disappear</u>.

3) The <u>quicker</u> the mark disappears, the <u>quicker</u> the reaction.

4) This only works for reactions where the initial solution is rather <u>see-through</u>.

5) The result is very <u>subjective</u> — <u>different people</u> might not agree
   over the <u>exact</u> point when the mark 'disappears'.

A precipitate is a solid formed in a reaction between solutions

Sadly, measuring the rate of a reaction using precipitation is quite subjective — two people generally
don't have the same view of when the marker disappears. The two methods you'll meet on the next
page are much more accurate, but they're only useful if you've got a reaction that produces a gas...

Measuring Rates of Reaction

2) *Change in Mass* (Usually Gas Given Off)

1) Measuring the speed of a reaction that <u>produces a gas</u> can be carried out on a <u>mass balance</u>.

2) As the gas is released the mass <u>disappearing</u> is easily measured on the balance.

3) The <u>quicker</u> the reading on the balance <u>drops</u>, the <u>faster</u> the reaction.

4) <u>Rate of reaction graphs</u> are particularly easy to plot using the results from this method.

5) This is the <u>most accurate</u> of the three methods described because the mass balance is very accurate. But it has the <u>disadvantage</u> of releasing the gas straight into the room.

3) The **Volume** of Gas Given Off

1) This involves the use of a <u>gas syringe</u> to measure the <u>volume</u> of gas given off.

2) The <u>more</u> gas given off during a given <u>time interval</u>, the <u>faster</u> the reaction.

3) A graph of <u>gas volume</u> against <u>time elapsed</u> could be plotted to give a rate of reaction graph.

4) Gas syringes usually give volumes accurate to the <u>nearest millilitre</u>, so they're quite accurate. You have to be quite careful though — if the reaction is too <u>vigorous</u>, you can easily blow the plunger out of the end of the syringe.

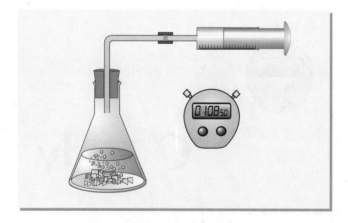

Each of these methods has pros and cons

The mass balance method is only accurate as long as the flask isn't too hot, otherwise you lose mass by <u>evaporation</u> as well as in the reaction. The first method <u>isn't</u> very accurate, but if you're not producing a gas you can't use either of the other two. Ah well.

Rate of Reaction Experiments

Remember: Any reaction can be used to investigate any of the four factors that affect the rate.
The next four pages illustrate four important reactions, but only one factor is considered for each.
But you can just as easily use, say, the marble chips/acid reaction to test the effect of temperature instead.

1) Reaction of **Hydrochloric Acid** and **Marble Chips**

This experiment is often used to demonstrate the effect of breaking the solid up into small bits.

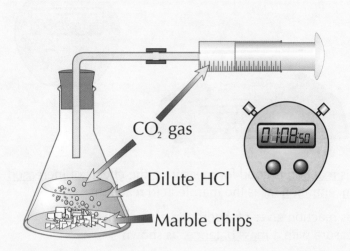

CO_2 gas

Dilute HCl

Marble chips

1) Measure the volume of gas evolved with a gas syringe and take readings at regular intervals.

2) Make a table of readings and plot them as a graph. You choose regular time intervals, and time goes on the x-axis and volume goes on the y-axis.

3) Repeat the experiment with exactly the same volume of acid, and exactly the same mass of marble chips, but with the marble more crunched up.

4) Then repeat with the same mass of powdered chalk instead of marble chips.

This Graph Shows the Effect of Using **Finer Particles of Solid**

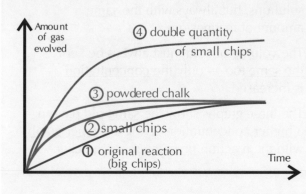

Amount of gas evolved

④ double quantity of small chips

③ powdered chalk

② small chips

① original reaction (big chips)

Time

1) Using finer particles means that the marble has a larger surface area.

2) A larger surface area causes more frequent collisions (see page 157) so the rate of reaction is faster.

3) Line 4 shows the reaction if a greater mass of small marble chips is added.
The extra surface area gives a quicker reaction and there is also more gas evolved overall.

Rate of Reaction Experiments

The reaction of <u>magnesium metal</u> with <u>dilute HCl</u> is often used to determine the effect of <u>concentration</u>.

2) Reaction of Magnesium Metal With Dilute HCl

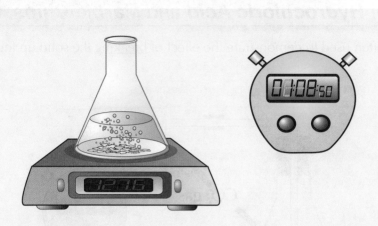

1) <u>This reaction</u> is good for measuring the effects of <u>increased concentration</u> (as is the marble/acid reaction).

2) This reaction gives off <u>hydrogen gas</u>, which we can measure with a <u>mass balance</u>, as shown.

3) In this experiment, <u>time</u> also goes on the <u>x-axis</u> and <u>volume</u> goes on the <u>y-axis</u>.
(The other method is to use a gas syringe, see page 148.)

This Graph Shows the Effect of Using More Concentrated Acid Solutions

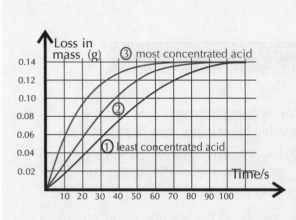

1) Take <u>readings</u> of mass at <u>regular</u> time intervals.

2) Put the results in a <u>table</u> and work out the <u>loss in mass</u> for each reading. <u>Plot a graph</u>.

3) <u>Repeat</u> with <u>more concentrated</u> acid solutions, but always with the <u>same</u> amount of magnesium.

4) The <u>volume</u> of acid must always be kept <u>the same</u> too — only the <u>concentration</u> is increased.

5) The three graphs show the <u>same</u> old pattern — a <u>higher</u> concentration giving a <u>steeper graph</u>, with the reaction <u>finishing</u> much quicker.

Rate of Reaction Experiments

The effect of <u>temperature</u> on the rate of a reaction can be measured using a <u>precipitation</u> reaction.

3) *Sodium Thiosulfate* and *HCl* Produce a *Cloudy Precipitate*

1) These two chemicals are both <u>clear solutions</u>.

2) They react together to form a <u>yellow precipitate</u> of <u>sulfur</u>.

3) The experiment involves watching a black mark <u>disappear</u> through the <u>cloudy sulfur</u> and <u>timing</u> how long it takes to go.

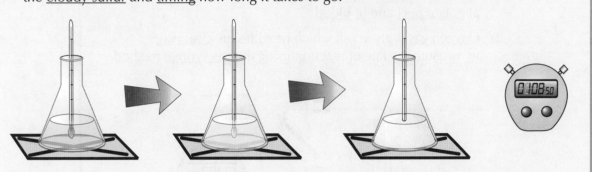

4) The reaction can be <u>repeated</u> for solutions at different <u>temperatures</u>. In practice, that's quite hard to do accurately and safely (it's not a good idea to heat an acid directly). The best way to do it is to use a <u>water bath</u> to heat both solutions to the right temperature <u>before you mix them</u>.

5) The <u>depth</u> of liquid must be kept the <u>same</u> each time, of course.

6) The results will of course show that the <u>higher</u> the temperature the <u>quicker</u> the reaction and therefore the <u>less time</u> it takes for the mark to <u>disappear</u>. These are typical results:

Temperature (°C)	20	25	30	35	40
Time taken for mark to disappear (s)	193	151	112	87	52

This reaction can <u>also</u> be used to test the effects of <u>concentration</u>.

This reaction <u>doesn't</u> give a set of graphs. All you get is a set of <u>readings</u> of how long it took till the mark disappeared for each temperature.

Rate of Reaction Experiments

Good news — this is the last rate experiment. This one looks at how a <u>catalyst</u> affects rate of reaction.

4) The **Decomposition** of **Hydrogen Peroxide**

This is a <u>good</u> reaction for showing the effect of different <u>catalysts</u>.
The decomposition of hydrogen peroxide is:

$$2H_2O_{2\,(aq)} \rightleftharpoons 2H_2O_{(l)} + O_{2\,(g)}$$

1) This is normally quite <u>slow</u> but a sprinkle of <u>manganese(IV) oxide catalyst</u> speeds it up no end. Other catalysts which work are found in: a) <u>potato peel</u> and b) <u>blood</u>.

2) <u>Oxygen gas</u> is given off, which provides an <u>ideal way</u> to measure the rate of reaction using the <u>gas syringe</u> method.

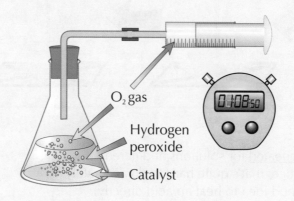

O₂ gas

Hydrogen peroxide

Catalyst

This Graph Shows the Effect of Using **Different Catalysts**

1) Same old graphs of course.

2) <u>Better</u> catalysts give a <u>quicker reaction</u>, which is shown by a <u>steeper graph</u> which levels off quickly.

3) This reaction can also be used to measure the effects of <u>temperature</u>, or of <u>concentration</u> of the H_2O_2 solution. The graphs will look just the same.

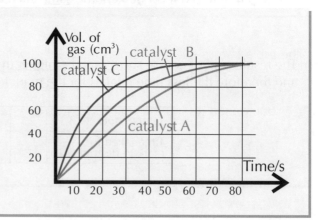

Blood is a catalyst? — eeurgh...

This stuff's about comparing those pretty rate of reaction graphs. They tell you the amount of <u>product</u> made (or reactant used up) and the <u>rate of reaction</u>. The <u>steeper</u> the curve, the <u>faster</u> the reaction.

Warm-Up and Exam Questions

Time to test your knowledge again. This time on the rates of chemical reactions. If you struggle with these questions and you don't feel up to speed, it's time to have another look at the last few pages.

Warm-Up Questions

1) Give an example of a reaction that happens very slowly, and one that is very fast.

2) Oxidation of lactose in milk makes it go 'sour'. How could this reaction be slowed down?

3) Give three ways of increasing the rate of a reaction between magnesium and sulfuric acid.

4) Describe one way of monitoring a reaction in which a gas is given off.

5) How would reducing the concentration of an acid affect the time taken for a piece of zinc to react with it?

Exam Questions

1   Set volumes of sodium thiosulfate and hydrochloric acid were reacted at different temperatures. The time taken for a black cross to be obscured by the sulfur precipitated was measured at each temperature. The results are shown in the table.

Temperature (°C)	Time (s)
55	6
36	11
24	17
16	27
9	40
5	51

(a) Give two variables that should be kept constant in this experiment.

(2 marks)

(b) Plot the results on a graph (with time on the x-axis) and draw a best-fit curve.

(2 marks)

(c) Describe the relationship illustrated by your graph.

(1 mark)

(d) Describe how the results would change if the sodium thiosulfate concentration was reduced.

(2 marks)

(e) Suggest how the results of the experiment could be made more reliable.

(1 mark)

2   The table shows the results of reactions between excess marble and 50 cm³ of 1 mol/dm³ hydrochloric acid.

Time (min)	Mass of flask A (g)	Mass of flask B (g)
0	121.6	121.6
1	120.3	119.8
2	119.7	119.2
3	119.4	119.1
4	119.2	119
5	119.1	119
6	119	119
7	119	

(a) Explain why the mass of the contents of the flasks decreased during the reaction.

(1 mark)

(b) Explain why the mass of each flask and its contents fell by the same total amount.

(1 mark)

(c) Suggest what conditions may have been different inside flask B.

(1 mark)

(d) In both reactions, the rate is fastest at the beginning. Suggest why.

(1 mark)

Rate of Reaction Data

You can learn a lot from a <u>rate of reaction</u> graph. Read on to explore this land of graphical fun...

Reaction Rate Graphs Show Rate of Reaction Data

Surface Area Affects Rate of Reaction

When marble chips are added to hydrochloric acid, CO_2 is given off.

In this experiment, 5 g of marble chips were added to hydrochloric acid, and the volume of CO_2 measured every 10 seconds. The results are plotted below. Line 1 is for <u>small chips</u> and line 2 is for <u>large chips</u>.

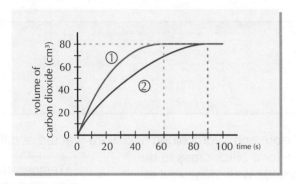

1) Both reactions finish (the line goes flat) when 80 cm³ of CO_2 are produced.

2) The <u>reaction time</u> for Reaction 1 is about 60 s, and for Reaction 2 it's about 90 s — <u>Reaction 1 is faster</u>.

3) Another way to tell the rate of reaction is to look at the <u>slope</u> of the graph — the <u>steeper</u> the graph, the <u>faster</u> the reaction. Reaction 1 is <u>faster</u> than Reaction 2 — the <u>slope of its graph is steeper</u>.

4) <u>Reaction 1</u> is <u>faster</u> because small chips have a <u>larger surface area</u> than the same mass of large chips.

Concentration Affects Rate of Reaction

This time, a piece of magnesium has been added to hydrochloric acid. The graphs show the volume of hydrogen produced when two <u>different concentrations</u> of acid are used.

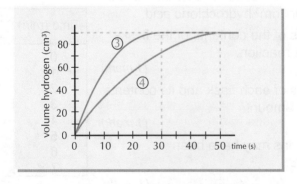

1) <u>Reaction 3</u> is <u>faster</u> than Reaction 4 — its <u>slope is steeper</u> (or use the fact that Reaction 3 takes about 30 s, and Reaction 4 about 50 s).

2) Since Reaction 3 is <u>faster</u>, it must use the <u>more concentrated</u> acid.

Rate Of Reaction Data

Temperature Affects Rate of Reaction

In this version of the experiment the size of the marble chips is the same but two <u>different temperatures</u> of acid are used.

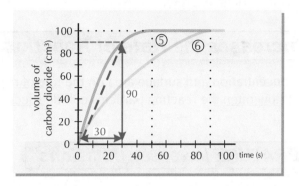

1) Both reactions finish when 100 cm³ of CO_2 have been produced.

2) Reaction 5 is faster (about 50 s, compared to Reaction 6's 90 s or so).

3) You can calculate the <u>rate of reaction</u> by calculating the slope of the line.

4) To find the <u>average rate</u> during the first 30 s, draw a line from the volume of CO_2 produced at the start to the volume produced at 30 s then find the slope of this line.

5) For Reaction 5, the slope is 90 ÷ 30 = 3.
   This means that you're getting <u>3 cm³ of CO_2 per second (3 cm³/s)</u>.

A Catalyst Affects Rate of Reaction

This graph shows the effect of adding a catalyst.
Line 7 is the <u>catalysed</u> reaction, line 8 is the <u>uncatalysed</u> reaction.

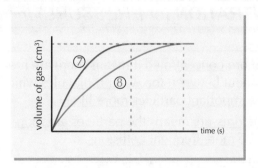

The graph tells you that the <u>catalyst speeds up the reaction</u> because line 7 is <u>steeper</u> than line 8 (but the <u>total volume of gas</u> produced is the <u>same</u> in each reaction).

My reactions slow down when it gets hot — I get sleepy...

When you're trying to work out the units of a reaction rate, take a look at the data you've got. If the data is in cm³ and seconds then the units will be cm³ per s (cm³/s). If the data is in grams and minutes then the units will be g per min (g/min). Come on, it's not so bad...

Collision Theory

Reaction rates are explained by collision theory. It's really simple.

1) Collision theory just says that the rate of a reaction simply depends on how often and how hard the reacting particles collide with each other.

2) The basic idea is that particles have to collide in order to react, and they have to collide hard enough (with enough energy).

More Collisions Increases the Rate of Reaction

The effects of temperature, concentration and surface area on the rate of reaction can be explained in terms of how often the reacting particles collide successfully.

1) HIGHER TEMPERATURE Increases Collisions

When the temperature is increased the particles all move quicker. If they're moving quicker, they're going to collide more often.

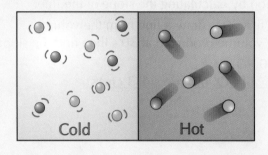

Cold   Hot

2) HIGHER CONCENTRATION (or PRESSURE) Increases Collisions

If a solution is made more concentrated it means there are more particles of reactant knocking about between the water molecules which makes collisions between the important particles more likely.

In a gas, increasing the pressure means the particles are more squashed up together so there will be more frequent collisions.

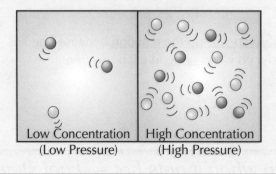

Low Concentration   High Concentration
(Low Pressure)   (High Pressure)

Collision Theory

3) *LARGER SURFACE AREA* Increases **Collisions**

If one of the reactants is a <u>solid</u> then <u>breaking it up</u> into <u>smaller</u> pieces will <u>increase the total surface area</u>. This means the particles around it in the solution will have <u>more area to work on</u>, so there'll be <u>more frequent collisions</u>.

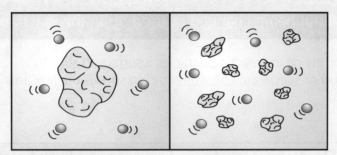

Small surface area   Big surface area

Faster collisions Increase the Rate of Reaction

1) <u>Higher temperature</u> also increases the <u>energy</u> of the collisions, because it makes all the particles <u>move faster</u>.

<u>Increasing</u> the <u>temperature</u> causes <u>faster collisions</u>.

2) Reactions <u>only happen</u> if the particles collide with <u>enough energy</u>.

3) The <u>minimum amount</u> of energy needed by the particles to react is known as the <u>activation energy</u>.

4) At a <u>higher temperature</u> there will be <u>more particles</u> colliding with <u>enough energy</u> to make the reaction happen.

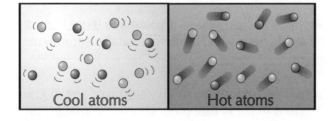

Cool atoms   Hot atoms

Chemical reactions depend on collisions

Once you've read everything on this page, the rates of reaction stuff should start making <u>a lot more sense</u> to you. The concept's fairly simple — the <u>more often</u> particles bump into each other, and the <u>harder</u> they hit when they do, the <u>faster</u> the reaction happens.

Catalysts

In industrial reactions, the main thing they're interested in is making a nice profit.
Catalysts are helpful for this — they can reduce costs and increase the amount of product.

Catalysts *Speed Up* Reactions

Many reactions can be speeded up by adding a catalyst.

> A catalyst is a substance which speeds up a reaction,
> without being changed or used up in the reaction.

A solid catalyst works by giving the reacting particles a surface to stick to.
This increases the number of successful collisions (and so speeds the reaction up).

Catalysts Help *Reduce Costs* in Industrial Reactions

1) Catalysts are very important for commercial reasons — most industrial reactions use them.

2) Catalysts increase the rate of the reaction, which saves a lot of money simply because the plant doesn't need to operate for as long to produce the same amount of stuff.

3) Alternatively, a catalyst will allow the reaction to work at a much lower temperature. That reduces the energy used up in the reaction (the energy cost), which is good for sustainable development (see page 133) and can save a lot of money too.

4) There are disadvantages to using catalysts, though.

5) They can be very expensive to buy, and often need to be removed from the product and cleaned. They never get used up in the reaction though, so once you've got them you can use them over and over again.

6) Different reactions use different catalysts, so if you make more than one product at your plant, you'll probably need to buy different catalysts for them.

7) Catalysts can be 'poisoned' by impurities, so they stop working, e.g. sulfur impurities can poison the iron catalyst used in the Haber process (used to make ammonia for fertilisers). That means you have to keep your reaction mixture very clean.

A big advantage of catalysts is that they can be used over and over

And they're not only used in industry... every useful chemical reaction in the human body is catalysed by a biological catalyst (an enzyme). If the reactions in the body were just left to their own devices, they'd take so long to happen, we couldn't exist. Quite handy then, these catalysts.

Warm-Up and Exam Questions

Here are some more questions to have a go at. If you can't do these ones then you won't be able to do the ones in the exams either. And you don't want that. If you're struggling, read the pages over again.

Warm-Up Questions

1) Explain how you would calculate the rate of reaction from a reaction rate graph.
2) Why does an increase in concentration of solutions increase the rate of a reaction?
3) According to collision theory, what must happen in order for two particles to react?
4) Give a definition of a catalyst.

Exam Questions

1   *In this question you will be assessed on the quality of your English,
the organisation of your ideas and your use of appropriate specialist vocabulary.*

Hydrogen and ethene react to form ethane. Nickel can be used as a catalyst for this reaction.

Using your knowledge of collision theory, suggest how the rate of this reaction can
be increased.

(6 marks)

2   The contact process is used in industry to produce sulfuric acid.
Vanadium oxide can be used to catalyse this process.

(a) Describe how the solid vanadium oxide catalyst used in the
contact process will help to speed up the rate of reaction.

(2 marks)

(b) Although catalysts are used widely in the chemical industry,
they do have some disadvantages.
Circle the statements below that are **disadvantages**
of using catalysts in industrial processes.

You have to use a new catalyst if you want to repeat a reaction.	Catalysts are expensive to buy.
The energy used up by the process will be reduced.	Different reactions require different catalysts so a different catalyst will have to be bought for every reaction that takes place.
Catalysts will only work at very high temperatures — creating the very high temperatures is expensive.	Catalysts will never run out.

(2 marks)

Energy Transfer in Reactions

Whenever chemical reactions occur <u>energy</u> is <u>transferred to</u> or <u>from</u> the <u>surroundings</u>.

In an **Exothermic** Reaction, Heat is **Given Out**

An <u>EXOTHERMIC reaction</u> is one which <u>transfers energy</u> to the surroundings, usually in the form of <u>heat</u> and usually shown by a <u>rise in temperature</u>.

1) **Burning** Fuels

The best example of an <u>exothermic</u> reaction is <u>burning fuels</u> — also called <u>COMBUSTION</u>. This gives out a lot of heat — it's very exothermic.

2) **Neutralisation** Reactions

<u>Neutralisation reactions</u> (acid + alkali) are also exothermic — see page 166.

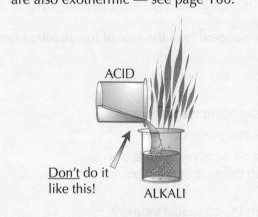

ACID

<u>Don't</u> do it like this!

ALKALI

3) **Oxidation** Reactions

Many <u>oxidation reactions</u> are exothermic. For example:

Adding sodium to water <u>produces heat</u>, so it must be <u>exothermic</u> — see page 101. The sodium emits <u>heat</u> and moves about on the surface of the water as it is oxidised.

Exothermic reactions have lots of <u>everyday uses</u>. For example, some <u>hand warmers</u> use the exothermic <u>oxidation of iron</u> in air (with a salt solution catalyst) to generate <u>heat</u>. <u>Self heating cans</u> of hot chocolate and coffee also rely on exothermic reactions between <u>chemicals</u> in their bases.

Energy Transfer in Reactions

*In an **Endothermic** Reaction, Heat is **Taken In***

An <u>ENDOTHERMIC reaction</u> is one which <u>takes in energy</u> from the surroundings, usually in the form of <u>heat</u> and is usually shown by a <u>fall in temperature</u>.

Endothermic reactions are much <u>less common</u>. <u>Thermal decompositions</u> are a good example:

Heat must be supplied to make calcium carbonate <u>decompose</u> to make calcium oxide.

$$CaCO_3 \rightarrow CaO + CO_2$$

Endothermic reactions also have everyday uses. For example, some <u>sports injury packs</u> use endothermic reactions — they <u>take in heat</u> and the pack becomes very <u>cold</u>. More <u>convenient</u> than carrying ice around.

Reversible Reactions** Can Be **Endothermic** and **Exothermic

In reversible reactions (see page 133), if the reaction is <u>endothermic</u> in <u>one direction</u>, it will be <u>exothermic</u> in the <u>other direction</u>. The <u>energy absorbed</u> by the endothermic reaction is <u>equal</u> to the <u>energy released</u> during the exothermic reaction.

A good example is the <u>thermal decomposition of hydrated copper sulfate</u>.

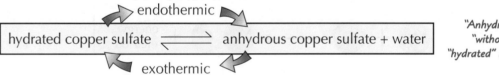

endothermic

hydrated copper sulfate ⇌ anhydrous copper sulfate + water

exothermic

"Anhydrous" just means "without water", and "hydrated" means "with water".

1) If you <u>heat blue hydrated</u> copper(II) sulfate crystals it drives the water off and leaves <u>white anhydrous</u> copper(II) sulfate powder. This is endothermic.

Water vapour

2) If you then <u>add</u> a couple of drops of <u>water</u> to the <u>white powder</u> you get the <u>blue crystals</u> back again. This is exothermic.

Right, so burning gives out heat — really...

This whole energy transfer thing is a fairly simple idea — don't be put off by the long words. Remember, "<u>exo-</u>" = <u>exit</u>, "<u>-thermic</u>" = <u>heat</u>, so an exothermic reaction is one that <u>gives out</u> heat. And "<u>endo-</u>" = erm... the other one. Okay, so there's no easy way to remember that one. Tough.

Energy

Whenever chemical reactions occur, there are changes in <u>energy</u>. Changes in energy during a chemical reaction can be explained by <u>making bonds</u> or <u>breaking bonds</u>.

*Energy Must Always be **Supplied** to **Break Bonds**...*
*...and Energy is Always **Released** When **Bonds Form***

1) During a chemical reaction, <u>old bonds</u> are <u>broken</u> and <u>new bonds</u> are <u>formed</u>.

2) Energy must be <u>supplied</u> to break <u>existing bonds</u> — so bond breaking is an <u>endothermic</u> process. Energy is <u>released</u> when new bonds are <u>formed</u> — so bond formation is an <u>exothermic</u> process.

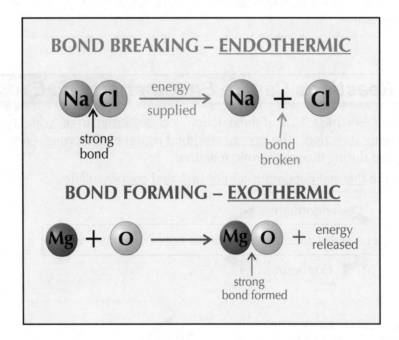

3) In an <u>endothermic</u> reaction, the energy <u>required</u> to break old bonds is <u>greater</u> than the energy <u>released</u> when <u>new bonds</u> are formed.

4) In an <u>exothermic</u> reaction, the energy <u>released</u> in bond formation is <u>greater</u> than the energy used in <u>breaking</u> old bonds.

Save energy — break fewer bonds...

I reckon that this <u>bond breaking = endothermic</u>, <u>bond forming = exothermic</u> thing is one of the trickiest things to get your head around. But don't panic — all you need to say is <u>FEXBEN</u> — 'forming = ex, breaking = en'. Well, OK that kinda sucks but I'd like to see you come up with something better.

Energy Level Diagrams and Temperature Change

Chemical reactions involve a change in energy. Energy level diagrams show this change.

Energy Level Diagrams Show if it's Exothermic or Endothermic

Exothermic

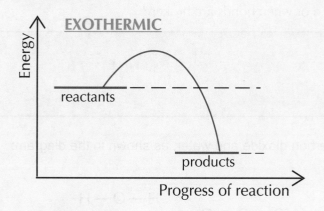

1) This shows an exothermic reaction — the products are at a lower energy than the reactants.
2) The difference in height represents the energy given out in the reaction.
3) The initial rise in the line represents the energy needed to break the old bonds.

Endothermic

1) This shows an endothermic reaction because the products are at a higher energy than the reactants.
2) The difference in height represents the energy taken in during the reaction.

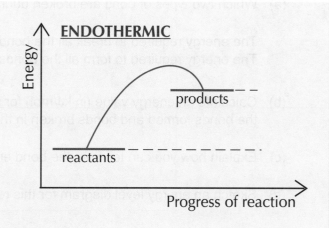

Temperature Change can be Measured

You can measure the amount of energy produced by a chemical reaction in solution by taking the temperature of the reactants (making sure they're the same), mixing them in a polystyrene cup and measuring the temperature of the solution at the end of the reaction.

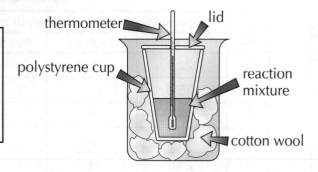

1) The biggest problem with temperature measurements is the amount of heat lost to the surroundings.
2) You can reduce it a bit by putting the polystyrene cup into a beaker of cotton wool to give more insulation, and putting a lid on the cup to reduce energy lost by evaporation.

Warm-Up and Exam Questions

Some of this stuff can be a bit tricky — have a go at these to make sure you've got it.

Warm-Up Questions

1) Are most oxidation reactions exothermic or endothermic?

2) An endothermic reaction happens when ammonium nitrate is dissolved in water. Predict how the temperature of the solution will change during the reaction.

3) Is energy released when bonds are formed or when bonds are broken?

Exam Questions

1   When methane burns in air it produces carbon dioxide and water, as shown in the diagram:

$$
\begin{array}{c}
\text{H} \\
| \\
\text{H}-\text{C}-\text{H} \\
| \\
\text{H}
\end{array}
+
\begin{array}{c}
\text{O}=\text{O} \\
\text{O}=\text{O}
\end{array}
\rightarrow
\text{O}=\text{C}=\text{O}
+
\begin{array}{c}
\text{H}-\text{O}-\text{H} \\
\text{H}-\text{O}-\text{H}
\end{array}
$$

(a)   Which two types of bond are broken during the reaction?

(1 mark)

The energy required to break all the bonds in this reaction is 2644 kJ/mol.
The energy required to form all the bonds in this reaction is 3436 kJ/mol.

(b)   Calculate an energy value (in kJ/mol) for the difference between the bonds formed and bonds broken in this reaction.

(1 mark)

(c)   Explain how you can tell from the bond energies that the reaction is exothermic.

(1 mark)

(d)   Sketch an energy level diagram for this reaction.

(2 marks)

2   A student added hydrochloric acid to sodium hydroxide. He measured the temperature of the reaction mixture over the first 5 seconds and recorded his results in the table.

Time (s)	Temperature of the reaction mixture (°C)		
	1st run	2nd run	Average
0	22.0	22.0	
1	25.6	24.4	
2	28.3	28.1	
3	29.0	28.6	
4	28.8	28.8	
5	28.3	28.7	

(a)   Complete the table by calculating the average temperature of the reaction mixture during the two runs.

(2 marks)

(b)   Calculate the maximum average increase in temperature during the reaction.

(1 mark)

(c)   Is this reaction exothermic or endothermic? Explain your answer.

(1 mark)

Revision Summary for Section 5

Well, I don't think that was too bad, was it... There are loads of methods you can use to identify substances. Chemical synthesis is used by the chemical industry to produce the chemicals that we use everyday. Four things affect the rate of reactions, there are loads of ways to measure reaction rates and it's all explained by collision theory. Reactions can be endothermic or exothermic. And so on... Easy.

Well here are some more of those nice questions that you enjoy so much. If there are any you can't answer, go back to the appropriate page, do a bit more learning, then try again.

1) A student makes a solution of an unknown compound and adds a couple of drops of sodium hydroxide. He gets a white precipitate. He adds more sodium hydroxide and the precipitate dissolves.
 a) What positive ion is present?
 b) Write down an ionic equation for the formation of the white precipitate.

2) Iron(II) chloride forms a sludgy green precipitate with sodium hydroxide. Write down an ionic equation for this reaction.

3) Carbonates give off carbon dioxide when they're mixed with dilute acid. Write an ionic equation for the reaction between carbonates and dilute acid.

4) What's the test for sulfates?

5) Explain how paper chromatography can be used to analyse the dyes used in a brown sweet.

6) Briefly describe how gas chromatography works.

7) What are 'batch production' and 'continuous production'?

8) Give an example of one type of chemical that is produced on a small scale.

9) It can take 12 years and about £900 million to bring a new drug to market. Explain why.

10) What are the four factors that affect the rate of a reaction?

11) Give two reasons why it is important to control the rate of a chemical reaction in industry.

12) Describe three different ways of measuring the rate of a reaction.

13) A student carries out an experiment to measure the effect of surface area on the reaction between marble and hydrochloric acid. He measures the amount of gas given off at regular intervals.
 a) What factors must he keep constant for it to be a fair test?
 b)* He uses four samples for his experiment:
  Sample A – 10 g of powdered marble
  Sample B – 10 g of small marble chips
  Sample C – 10 g of large marble chips
  Sample D – 5 g of powdered marble
  Sketch a typical set of graphs for this experiment.

14) A piece of magnesium is added to a dilute solution of hydrochloric acid, and hydrogen gas is produced. The experiment is repeated with a more concentrated hydrochloric acid. How can you tell from the experiment which concentration of acid produces a faster rate of reaction?

15) Explain how higher temperature, higher concentration and larger surface area increase the frequency of successful collisions between particles.

16) What is an exothermic reaction? Give three examples.

17) The reaction to split ammonium chloride into ammonia and hydrogen chloride is endothermic. What can you say for certain about the reverse reaction?

18) a) Draw graphs showing energy change in endothermic and exothermic reactions.
 b) Explain how bond breaking and bond forming relate to these graphs.

19) How would you measure the temperature change during a reaction?

* Answer on page 276.

Acids and Alkalis

Testing the pH of a solution means using an <u>indicator</u> — and that means pretty <u>colours</u>...

The *pH Scale* Goes from *0 to 14*

1) The <u>pH scale</u> is a measure of how <u>acidic</u> or <u>alkaline</u> a solution is.
2) The <u>strongest acid</u> has <u>pH 0</u>. The <u>strongest alkali</u> has <u>pH 14</u>.
3) A <u>neutral</u> substance has <u>pH 7</u> (e.g. pure water).

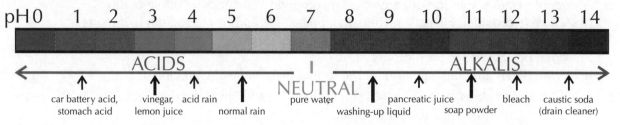

Indicators and *pH Meters* can be Used to Determine *pH*

1) Indicators contain a dye that <u>changes colour</u> depending on whether it's <u>above</u> or <u>below</u> a certain pH.
2) <u>Litmus paper</u> is an easy way to find out if a solution is acidic or alkaline — it turns <u>red</u> if the solution is <u>acidic</u> and <u>blue</u> if it's <u>alkaline</u>.
3) <u>Universal indicator</u> is a very useful <u>combination of dyes</u>, which gives the colours shown above. It's useful for <u>estimating</u> the pH of a solution.
4) <u>pH meters</u> can also be used to measure the pH of a substance. These usually consist of a <u>probe</u>, which is dipped into the substance, and a <u>meter</u>, which gives a reading of the pH.
5) pH meters are <u>more accurate</u> than indicators.

Acids and *Bases* Neutralise Each Other

> An <u>ACID</u> is a substance with a pH of less than 7. Acids form <u>H$^+$ ions</u> in <u>water</u>.
> A <u>BASE</u> is a substance with a pH of greater than 7.
> An <u>ALKALI</u> is a base that <u>dissolves in water</u>. Alkalis form <u>OH$^-$ ions</u> in <u>water</u>.
> So, <u>H$^+$</u> ions make solutions <u>acidic</u> and <u>OH$^-$</u> ions make them <u>alkaline</u>.

The reaction between acids and bases is called <u>neutralisation</u>:

$$acid + base \rightarrow salt + water$$

Neutralisation can also be seen in terms of <u>H$^+$</u> and <u>OH$^-$</u> ions like this:

$$H^+_{(aq)} + OH^-_{(aq)} \rightarrow H_2O_{(l)}$$

Hydrogen (H$^+$) ions react with hydroxide (OH$^-$) ions to produce water.

When an acid neutralises a base (or vice versa), the <u>products</u> are <u>neutral</u>, i.e. they have a <u>pH of 7</u>. An indicator can be used to show that a neutralisation reaction is over (Universal indicator will go green).

State Symbols Tell You What *Physical State* it's in

These are easy enough, <u>so make sure you know them</u> — especially aq (aqueous).

(s) — Solid	(l) — Liquid	(g) — Gas	(aq) — Dissolved in water

E.g. $2Mg_{(s)} + O_{2\,(g)} \rightarrow 2MgO_{(s)}$

Acids Reacting With Metals

There are loads of different salts out there. Some of them are made when <u>acids</u> react with <u>metals</u>.

Metals React with Acids to Give Salts

$$\boxed{\text{acid } + \text{ metal } \rightarrow \text{ salt } + \text{ hydrogen}}$$

That's written big because it's really useful. Here's the <u>typical experiment</u>:

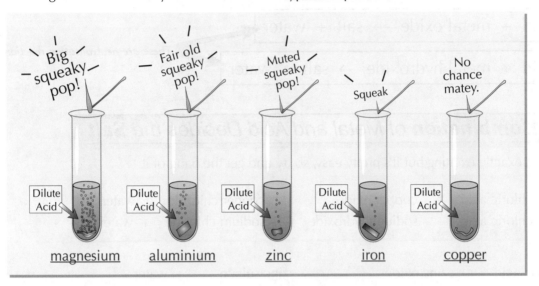

1) The more <u>reactive</u> the metal, the <u>faster</u> the reaction will go — very reactive metals (e.g. sodium) react <u>explosively</u>.

2) <u>Copper</u> does <u>not</u> react with dilute acids <u>at all</u> — because it's <u>less</u> reactive than <u>hydrogen</u>.

3) The <u>speed</u> of reaction is indicated by the <u>rate</u> at which the <u>bubbles</u> of hydrogen are given off.

4) The <u>hydrogen</u> is confirmed by the <u>burning splint test</u> giving the notorious 'squeaky pop'.

5) The <u>name</u> of the <u>salt</u> produced depends on which <u>metal</u> is used, and which <u>acid</u> is used:

Hydrochloric acid will always produce chloride salts:

$2HCl + Mg \rightarrow MgCl_2 + H_2$ (Magnesium chloride)
$6HCl + 2Al \rightarrow 2AlCl_3 + 3H_2$ (Aluminium chloride)
$2HCl + Zn \rightarrow ZnCl_2 + H_2$ (Zinc chloride)

Sulfuric acid will always produce sulfate salts:

$H_2SO_4 + Mg \rightarrow MgSO_4 + H_2$ (Magnesium sulfate)
$3H_2SO_4 + 2Al \rightarrow Al_2(SO_4)_3 + 3H_2$ (Aluminium sulfate)
$H_2SO_4 + Zn \rightarrow ZnSO_4 + H_2$ (Zinc sulfate)

Nitric acid produces nitrate salts when NEUTRALISED, but...

Nitric acid reacts fine with alkalis, to produce nitrates, but it can play silly devils with metals and produce nitrogen oxides instead, so we'll ignore it here.

Oxides, Hydroxides and Ammonia

I'm afraid there's more stuff on <u>neutralisation</u> reactions coming up...

Metal **Oxides** and Metal **Hydroxides** are **Bases**

1) Some <u>metal oxides</u> and <u>metal hydroxides</u> dissolve in <u>water</u>. These soluble compounds are <u>alkalis</u>.

2) Even bases that won't dissolve in water will still react with acids.

3) So, all <u>metal oxides</u> and <u>metal hydroxides</u> react with <u>acids</u> to form a <u>salt</u> and <u>water</u>.

acid + metal oxide → salt + water

acid + metal hydroxide → salt + water

These are neutralisation reactions.

The **Combination** of Metal and Acid Decides the **Salt**

This isn't exactly exciting but it's pretty easy, so try and get the hang of it:

hydrochloric acid + copper oxide → copper chloride + water
hydrochloric acid + sodium hydroxide → sodium chloride + water

sulfuric acid + zinc oxide → zinc sulfate + water
sulfuric acid + calcium hydroxide → calcium sulfate + water

nitric acid + magnesium oxide → magnesium nitrate + water
nitric acid + potassium hydroxide → potassium nitrate + water

The symbol equations are all pretty much the same. Here are two of them:

$$H_2SO_{4\ (aq)} + ZnO_{(s)} \rightarrow ZnSO_{4\ (aq)} + H_2O_{(l)}$$
$$HNO_{3\ (aq)} + KOH_{(aq)} \rightarrow KNO_{3\ (aq)} + H_2O_{(l)}$$

Metal **Carbonates** Give **Salt + Water + Carbon Dioxide**

More gripping reactions involving acids. At least there are some <u>bubbles</u> involved here.

Acid + Metal Carbonate → Salt + Water + Carbon Dioxide

The reaction is the same as the neutralisation reaction above EXCEPT that <u>carbon dioxide</u> is given off too.

Ammonia Can be **Neutralised** with HNO_3 to Make **Fertiliser**

<u>Ammonia</u> dissolves in water to make an <u>alkaline solution</u>.
When it reacts with <u>nitric acid</u>, you get a <u>neutral salt</u> — <u>ammonium nitrate</u>:

$$NH_{3\ (aq)} + HNO_{3\ (aq)} \rightarrow NH_4NO_{3\ (aq)}$$
ammonia + nitric acid → ammonium nitrate

This is a bit different from most neutralisation reactions because there's <u>NO WATER</u> produced — just the ammonium salt.

<u>Ammonium nitrate</u> is an especially good fertiliser because it has <u>nitrogen</u> from <u>two sources</u>, the ammonia and the nitric acid. Kind of a <u>double dose</u>. Plants need nitrogen to make <u>proteins</u>.

Making Soluble Salts

Most chlorides, sulfates and nitrates are soluble in water (the main exceptions are lead chloride, lead sulfate and silver chloride). Most oxides and hydroxides are insoluble in water.

The method you use to make a soluble salt depends on whether the base you use is soluble or not.

Making **Soluble Salts** Using a **Metal** or an **Insoluble Base**

1) You need to pick the right acid, plus a metal or an insoluble base (a metal oxide or metal hydroxide). E.g. if you want to make copper chloride, mix hydrochloric acid and copper oxide.

Remember some metals are unreactive and others are too reactive to use for this reaction (see page 167).

$$CuO_{(s)} + 2HCl_{(aq)} \rightarrow CuCl_{2\,(aq)} + H_2O_{(l)}$$

2) You add the metal, metal oxide or hydroxide to the acid — the solid will dissolve in the acid as it reacts. You will know when all the acid has been neutralised because the excess solid will just sink to the bottom of the flask.

3) Then filter out the excess metal, metal oxide or metal hydroxide to get the salt solution. To get pure, solid crystals of the salt, evaporate some of the water (to make the solution more concentrated) and then leave the rest to evaporate very slowly. This is called crystallisation.

filter paper

filter funnel

Making **Soluble Salts** Using an **Alkali**

1) You can't use the method above with alkalis (soluble bases) like sodium, potassium or ammonium hydroxides, because you can't tell whether the reaction has finished — you can't just add an excess to the acid and filter out what's left.

2) You have to add exactly the right amount of alkali to just neutralise the acid — you need to use an indicator (see page 166) to show when the reaction's finished. Then repeat using exactly the same volumes of alkali and acid so the salt isn't contaminated with indicator.

3) Then just evaporate off the water to crystallise the salt as normal.

Make sure you pick the right method...

If you're making a soluble salt, you need to think carefully about what chemicals you'd need to use to get the salt you want and what method you'd use. If you're making the salt from an acid and a soluble reactant then you need to know exactly how much of the reactant to use. If you're making the salt using an insoluble reactant then you add a load to the acid and filter out any that's left at the end of the reaction. In both cases you're left with only a salt and water. Evaporate, crystallise, done.

Making Insoluble Salts

That last page was all about making <u>soluble salts</u>. This one's about making <u>insoluble salts</u>.

*Making **Insoluble** Salts — **Precipitation** Reactions*

1) If the salt you want to make is <u>insoluble</u>, you can use a <u>precipitation reaction</u>.

2) You just need to pick <u>two solutions</u> that contain the <u>ions</u> you need. E.g. to make <u>lead chloride</u> you need a solution which contains <u>lead ions</u> and one which contains <u>chloride ions</u>. So you can mix <u>lead nitrate solution</u> (most nitrates are soluble) with <u>sodium chloride solution</u> (all group 1 compounds are soluble).

$$\text{E.g.} \quad Pb(NO_3)_{2\,(aq)} + 2NaCl_{(aq)} \longrightarrow PbCl_{2\,(s)} + 2NaNO_{3\,(aq)}$$

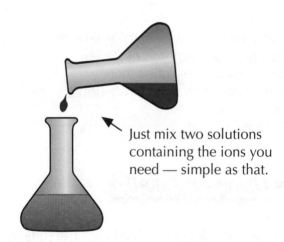

Just mix two solutions containing the ions you need — simple as that.

3) Once the salt has precipitated out (and is lying at the bottom of your flask), all you have to do is <u>filter</u> it from the solution, <u>wash</u> it and then <u>dry</u> it on filter paper.

4) <u>Precipitation reactions</u> can be used to remove <u>poisonous ions</u> (e.g. lead) from <u>drinking water</u>. <u>Calcium</u> and <u>magnesium</u> ions can also be removed from water this way — they make water "<u>hard</u>", which stops soap lathering properly. Another use of precipitation is in <u>treating effluent</u> (sewage) — again, <u>unwanted ions</u> can be removed.

Get two beakers, mix 'em together — job's a good'un...

All this stuff about insoluble and soluble salts can be a bit confusing at first — but try not to worry too much about it. If you think this method through <u>step-by-step</u> you'll get to grips with it in no time. It always helps to <u>see</u> reactions happening <u>first hand</u> — so if you can bully your teacher into letting you do this reaction in the lab then it'll be much easier for you to understand it.

Warm-Up and Exam Questions

Now try these questions — you're less likely to get a nasty surprise come exam time if you do.

Warm-Up Questions

1) What name is given to the type of reaction in which an acid reacts with a base?
2) Which two substances are formed when an acid reacts with a metal such as zinc?
3) Which two substances are formed when nitric acid reacts with copper oxide?
4) Explain what you would do to make a dry sample of a soluble salt from an insoluble base.
5) Write down the word equation for the precipitation reaction between barium chloride and sodium sulfate.

Exam Questions

1   An experiment was carried out in which sodium hydroxide solution was added, 2 cm^3 at a time, to 10 cm^3 of sulfuric acid. The pH was estimated after each addition using universal indicator paper.

The results are shown in the table.

Volume of sodium hydroxide added (cm^3)	pH
0	1
2	1
4	2
6	4
8	12
10	13
12	13

(a)   Plot the results on a graph, with pH on the vertical axis and volume of sodium hydroxide added on the horizontal axis.

Draw a best fit curve.

(2 marks)

(b)   Estimate the volume of sodium hydroxide needed to neutralise the acid.

(1 mark)

(c)   How do the results show that sulfuric acid is a strong acid?

(1 mark)

(d)   Name the salt formed in the reaction.

(1 mark)

2   Jenny wanted to make a dry sample of silver chloride, AgCl, by precipitation.

(a)   What property must a salt have to be made by precipitation?

(1 mark)

(b)   Jenny looked up the solubilities of some compounds she might use.

Write down a word equation using substances from the table that she could use to make silver chloride by precipitation.

(1 mark)

Compound	Formula	Solubility
silver oxide	Ag_2O	insoluble
silver nitrate	$AgNO_3$	soluble
silver carbonate	$AgCO_3$	insoluble
sulfuric acid	H_2SO_4	soluble
nitric acid	HNO_3	soluble
hydrochloric acid	HCl	soluble

(c)   Outline the steps needed to give a pure dry sample of silver chloride after mixing the solutions.

(3 marks)

Electrolysis

Examiners love <u>electrolysis</u>. It's just a shame that no one else does.

Electrolysis Means "Splitting Up with Electricity"

1) If you pass an <u>electric current</u> through an <u>ionic substance</u> that's <u>molten</u> or in <u>solution</u>, it breaks down into the <u>elements</u> it's made of. This is called <u>electrolysis</u>.

2) It requires a <u>liquid</u> to <u>conduct</u> the <u>electricity</u>, called the <u>electrolyte</u>.

3) Electrolytes contain <u>free ions</u> — they're usually the <u>molten</u> or <u>dissolved ionic substance</u>.

NaCl dissolved

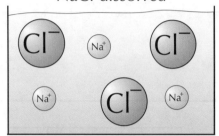

Molten NaCl

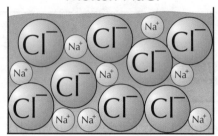

4) In either case it's the <u>free ions</u> which <u>conduct</u> the electricity and allow the whole thing to work.

5) For an electrical circuit to be complete, there's got to be a <u>flow of electrons</u>. <u>Electrons</u> are taken <u>away from</u> ions at the <u>positive electrode</u> and <u>given to</u> other ions at the <u>negative electrode</u>. As ions gain or lose electrons they become atoms or molecules and are released.

Electrolysis Reactions Involve **Oxidation** and **Reduction**

1) In Core Science you learnt about <u>reduction</u> involving the <u>loss of oxygen</u>. However...

2) <u>Reduction</u> is also a <u>gain of electrons</u>.

3) On the other hand, <u>oxidation</u> is a gain of oxygen or a <u>loss of electrons</u>.

4) So "reduction" and "oxidation" don't have to involve <u>oxygen</u>.

5) Electrolysis <u>ALWAYS</u> involves an oxidation and a reduction.

<u>O</u>xidation
<u>I</u>s
<u>L</u>oss of electrons

<u>R</u>eduction
<u>I</u>s
<u>G</u>ain of electrons

Remember it as OIL RIG.

Electrolysis needs a liquid to conduct the electricity

Before you electrolyse a substance it has to be <u>liquid</u>. This allows the ions to <u>move</u> towards the <u>positive</u> or <u>negative</u> electrode. The next few pages are about the electrolysis of different substances.

Electrolysis of Lead Bromide and Reactivity

Molten lead bromide can be broken down by electrolysis. You end up with lead and bromine.

The *Electrolysis* of Molten *Lead Bromide*

When a salt (e.g. lead bromide) is molten it will conduct electricity.

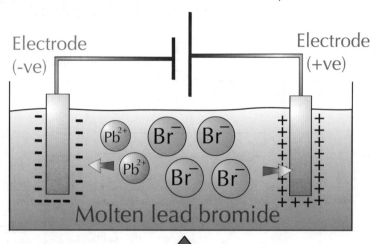

+ve ions are attracted
to the –ve electrode.
Here they gain
electrons (reduction).

–ve ions are attracted
to the +ve electrode.
Here they lose
electrons (oxidation).

Lead is produced
at the –ve electrode.

Bromine is produced
at the +ve electrode.

1) At the –ve electrode, one lead ion accepts two electrons to become one lead atom.

2) At the +ve electrode, two bromide ions lose one electron each
and become one bromine molecule.

As well as molten substances you can also electrolyse solutions. But first, a bit more about the products...

Reactivity Affects the *Products* Formed by *Electrolysis*

1) Sometimes there are more than two free ions in the electrolyte.
For example, if a salt is dissolved in water there will also be some H^+ and OH^- ions.

2) At the negative electrode, if metal ions and H^+ ions are present, the metal ions will stay in solution
if the metal is more reactive than hydrogen. This is because the more reactive an element, the
keener it is to stay as ions. So, hydrogen will be produced unless the metal is less reactive than it.

3) At the positive electrode, if OH^- and halide ions (Cl^-, Br^-, I^-) are present then molecules of chlorine,
bromine or iodine will be formed. If no halide is present, then oxygen will be formed.

So, lead bromide splits into lead and bromine — I know, it's tricky

The electrolysis of lead bromide is pretty simple — bromine ions are oxidised so bromine is produced
at the positive electrode and lead ions are reduced so lead is produced at the negative electrode.

Electrolysis of Sodium Chloride

You need to know about the electrolysis of salt (sodium chloride) solution. Get learning...

The **Electrolysis** of **Sodium Chloride Solution**

When common salt (sodium chloride) is dissolved in water and electrolysed,
it produces three useful products — <u>hydrogen</u>, <u>chlorine</u> and <u>sodium hydroxide</u>.

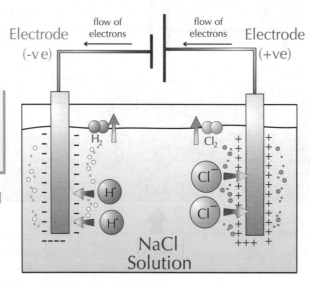

H⁺ ions are released from the water.

+ve ions are attracted to the –ve electrode. Here they <u>gain</u> <u>electrons</u> (reduction).

<u>Hydrogen</u> is produced at the <u>–ve electrode</u>.

–ve ions are attracted to the +ve electrode. Here they <u>lose</u> <u>electrons</u> (oxidation).

Chlorine is produced at the <u>+ve electrode</u>.

1) At the <u>negative electrode</u>, two hydrogen ions accept two electrons to become <u>one hydrogen molecule</u>.

2) At the <u>positive electrode</u>, two chloride (Cl^-) ions lose their electrons and become <u>one chlorine molecule</u>.

3) The <u>sodium ions</u> stay in solution because they're <u>more reactive</u> than hydrogen. <u>Hydroxide ions</u> from water are also left behind. This means that <u>sodium hydroxide</u> (NaOH) is left in the solution.

The **Half-Equations** — *Make Sure the Electrons Balance*

1) Half equations show the reactions at the electrodes. The main thing is to make sure the <u>number of electrons</u> is the <u>same</u> for <u>both half-equations</u>.

2) For the electrolysis of sodium chloride the half-equations are:

You need to make sure the atoms are balanced too.

Negative Electrode: $2H^+ + 2e^- \rightarrow H_2$
Positive Electrode: $2Cl^- \rightarrow Cl_2 + 2e^-$
or $2Cl^- - 2e^- \rightarrow Cl_2$

For the electrolysis of molten lead bromide (previous page) the half equations would be:
$Pb^{2+} + 2e^- \rightarrow Pb$
and $2Br^- \rightarrow Br_2 + 2e^-$

Products from the **Electrolysis of Sodium Chloride Solution**

The products of the electrolysis of sodium chloride solution are pretty useful in <u>industry</u>.

1) Chlorine has many uses, e.g. in the production of <u>bleach</u> and <u>plastics</u>.

2) Sodium hydroxide is a very strong <u>alkali</u> and is used <u>widely</u> in the <u>chemical industry</u>, e.g. to make <u>soap</u>.

Electrolysis of Aluminium

OK — one more example to learn about. This is the electrolysis of <u>aluminium oxide</u>.

Electrolysis is Used to Remove Aluminium From its Ore

1) Aluminium's a very <u>abundant</u> metal, but it is always found naturally in <u>compounds</u>.

2) Its main ore is <u>bauxite</u>, and after mining and purifying, a <u>white powder</u> is left.

3) This is <u>pure</u> aluminium oxide, Al_2O_3.

4) The <u>aluminium</u> has to be extracted from this using <u>electrolysis</u>.

Cryolite is Used to Lower the Temperature (and Costs)

1) Al_2O_3 has a very <u>high melting point</u> of over <u>2000 °C</u> — so melting it would be very <u>expensive</u>.

2) <u>Instead</u> the aluminium oxide is <u>dissolved</u> in <u>molten cryolite</u> (a less common ore of aluminium).

3) This brings the <u>temperature down</u> to about <u>900 °C</u>, which makes it much <u>cheaper</u> and <u>easier</u>.

4) The <u>electrodes</u> are made of <u>carbon</u> (graphite), a good conductor of electricity (see page 115).

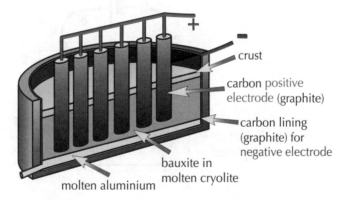

crust

carbon positive electrode (graphite)

carbon lining (graphite) for negative electrode

bauxite in molten cryolite

molten aluminium

Negative Electrode: $Al^{3+} + 3e^- \rightarrow Al$	Positive Electrode: $2O^{2-} \rightarrow O_2 + 4e^-$

5) <u>Aluminium</u> forms at the <u>negative electrode</u> and <u>oxygen</u> forms at the <u>positive electrode</u>.

6) The <u>oxygen</u> then reacts with the <u>carbon</u> in the electrode to produce <u>carbon dioxide</u>. This means that the <u>positive electrodes</u> gradually get 'eaten away' and have to be <u>replaced</u> every now and again.

It's all about lowering the cost...

The electrolysis of aluminium oxide may look <u>a bit different</u> to the examples on the other pages but don't be fooled. It's the same story — the positive aluminium ions go to the <u>negative</u> electrode and the negative oxygen ions are attracted to the <u>positive</u> electrode. Only one more page on electrolysis to go...

Electroplating

Electroplating coats one metal onto the surface of another. It's really useful...

Electroplating Uses Electrolysis

1) Electroplating uses electrolysis to <u>coat</u> the <u>surface of one metal</u> with <u>another metal</u>, e.g. you might want to electroplate silver onto a brass cup to make it look nice.

2) The <u>negative electrode</u> is the <u>metal object</u> you want to plate and the <u>positive electrode</u> is the <u>pure metal</u> you want it to be plated with. You also need the <u>electrolyte</u> to contain <u>ions</u> of the <u>plating metal</u>. (The ions that plate the metal object come from the solution, while the positive electrode keeps the solution 'topped up'.)

<u>Example</u>: To electroplate <u>silver</u> onto a <u>brass cup</u>, you'd make the <u>brass cup</u> the negative electrode (to attract the positive silver ions), a lump of <u>pure silver</u> the positive electrode and dip them in a solution of <u>silver ions</u>, e.g. silver nitrate.

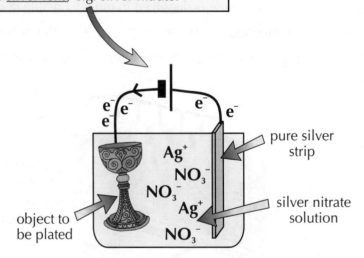

object to be plated

pure silver strip

silver nitrate solution

3) There are lots of different <u>uses</u> for electroplating:

• <u>Decoration</u>: <u>Silver</u> is <u>attractive</u>, but very <u>expensive</u>. It's much <u>cheaper</u> to plate a brass cup with silver, than it is to make the cup out of solid silver — but it looks just as <u>pretty</u>.

• <u>Conduction</u>: Metals like <u>copper</u> conduct <u>electricity</u> well — because of this they're often used to plate metals for <u>electronic circuits</u> and <u>computers</u>.

Silver electroplated text is worth a fortune...

There are loads of metals you can use for electroplating, but two really common ones are silver and copper. The tricky bit is remembering that the metal <u>object you want to plate</u> is the <u>negative electrode</u> and the <u>metal</u> you're plating it with is the <u>positive electrode</u>.

Warm-Up and Exam Questions

It's question time again. You know the drill. Off you go...

Warm-Up Questions

1) What state must an ionic compound be in if it's to be used as an electrolyte?
2) In terms of electrons, what is meant by the terms oxidation and reduction?
3) At which electrode are metals deposited during electrolysis?
4) During the electrolysis of molten lead bromide, which gas is produced at the positive electrode?
5) A metal object requires electroplating. Which electrode should it be used as?

Exam Questions

1   When sodium chloride solution is electrolysed a gas is produced at each electrode.
   (a) (i)   What is the name of the gas produced at the negative electrode?

(1 mark)

       (ii) State the half equation for the reaction at the negative electrode.

(1 mark)

   (b) (i)   What is the name of the gas produced at the positive electrode?

(1 mark)

       (ii) State the half equation for the reaction at the positive electrode.

(1 mark)

       (iii) Suggest one use for the gas produced at the positive electrode.

(1 mark)

   (c) Explain why sodium hydroxide is left in solution at the end of the reaction.

(3 marks)

2   The diagram shows a cell used to extract aluminium from aluminium oxide.

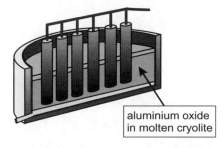

aluminium oxide
in molten cryolite

   (a) Explain why the aluminium oxide is dissolved in molten cryolite.

(2 marks)

   (b) Complete the half-equation below for the reaction at the negative electrode.

       _____ + 3e⁻ → _____

(1 mark)

   (c) The positive electrode is made out of carbon. Explain why it will need to be
       replaced over time.

(3 marks)

SECTION 6 — ACID REACTIONS, SALTS AND ELECTROLYSIS

Revision Summary for Section 6

Bleeuurgh... Who decided to make a section covering acids and bases <u>and</u> electrolysis — not the two nicest topics in chemistry in my view. Saying that, you still need to make sure that you can answer all of these fiendishly tricky* revision summary questions before you move on. You wouldn't want to lose marks in the exam just coz you didn't fancy learning about acids now, would you...

1) What does the pH scale show?

2) Name two ways of measuring the pH of a substance.

3) What type of ions are always present in a) acids and   b) alkalis?

4) What is neutralisation? Write down the general equation for neutralisation in terms of ions.

5) Write down the state symbol that means 'dissolved in water'.

6) What is the general equation for reacting an acid with a metal?

7) Name a metal that doesn't react at all with dilute acids.

8) What type of salts do hydrochloric acid and sulfuric acid produce?

9) What type of reaction is "acid + metal oxide", or "acid + metal hydroxide"?

10) Write a balanced symbol equation for the reaction between ammonia and nitric acid.
    What is the product of this reaction useful for?

11) Suggest a suitable acid and a suitable metal oxide/hydroxide to mix to form the following salts.
    a) copper chloride   b) calcium nitrate   c) zinc sulfate
    d) magnesium nitrate   e) sodium sulfate   f) potassium chloride

12) Iron chloride can made by mixing iron hydroxide (an insoluble base) with hydrochloric acid.
    Describe the method you would use to produce pure, solid iron chloride in the lab.

13) How can you tell when a neutralisation reaction is complete if both the base and the salt are soluble in water?

14) Give a practical use of precipitation reactions.

15) What is electrolysis? Explain why only liquids can be electrolysed.

16) Give one industrial use of sodium hydroxide.

17) Give two different uses of electroplating.

*Disclaimer: These revision summary questions are not really fiendishly tricky, more like a mild irritant.

Speed and Velocity

Speed and velocity aren't the same thing, you know. There's more to velocity than meets the eye.

Speed and Velocity are Both How Fast you're going

1) Speed and velocity are both measured in m/s (or km/h or mph).

2) They both simply say how fast you're going, but there's a subtle difference between them:

> **SPEED is just how fast you're going (e.g. 30 mph or 20 m/s) with no regard to the direction.**

> **VELOCITY however must also have the DIRECTION specified, e.g. 30 mph north or 20 m/s, 060°. The distance in a particular direction is called the DISPLACEMENT.**

3) Velocity and displacement are vector quantities — they have magnitude (size) and direction.

Speed, Distance and Time — the Formula:

Here's one of the most well known equations in physics...

$$\text{Speed} = \frac{\text{Distance}}{\text{Time}}$$

Example

> A cat skulks 20 m in 35 s. Find:
> a) its speed,   b) how long it takes to skulk 75 m.
>
> ANSWER: Using the formula triangle:
> a) s = d ÷ t = 20 ÷ 35 = 0.57 m/s
> b) t = d ÷ s = 75 ÷ 0.57 = 132 s = 2 min 12 s

A lot of the time we tend to use the words "speed" and "velocity" interchangeably.
But if you're talking about velocity, don't forget to state a direction.

Distance-Time Graphs

Distance-time graphs can tell you a lot about an object's movement.

Distance-Time Graphs

These are a very nifty way of describing something travelling through time and space:

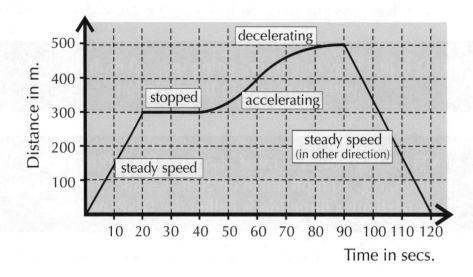

Very Important Notes:

1) <u>Gradient = speed</u>.

2) <u>Flat sections</u> are where it's <u>stopped</u>.

3) The <u>steeper</u> the graph, the <u>faster</u> it's going.

4) <u>Downhill</u> sections mean it's <u>going back</u> toward its starting point.

5) <u>Curves</u> represent <u>acceleration</u> or <u>deceleration</u>.

6) A <u>steepening</u> curve means it's <u>speeding up</u> (increasing gradient).

7) A <u>levelling off</u> curve means it's <u>slowing down</u> (decreasing gradient).

The *Gradient* of a *Distance-Time Graph* is the *Speed*

For example, the <u>speed</u> of the <u>return</u> section of the graph is:

$$\underline{Speed} = \underline{gradient} = \frac{vertical}{horizontal} = \frac{500}{30} = \underline{16.7 \ m/s}$$

This is just the speed equation (p. 179).

Don't forget that you have to use the <u>scales</u> of the axes to work out the gradient.
<u>Don't</u> measure in <u>cm</u>.

Acceleration

If something is <u>speeding up</u>, we say that it is <u>accelerating</u>.

Acceleration is How Quickly Velocity is Changing

Acceleration is <u>definitely not</u> the same as <u>velocity</u> or <u>speed</u> (see page 179).

> 1) Acceleration is <u>how quickly</u> the velocity is <u>changing</u>.
> 2) This change in velocity can be a <u>CHANGE IN SPEED</u> or a <u>CHANGE IN DIRECTION</u> or <u>both</u>.
>   (You only have to worry about the change in speed bit for calculations.)
> 3) <u>Deceleration</u> is just negative <u>acceleration</u>
>   (if something slows down, the change in velocity is negative).

<u>But</u>, acceleration is a <u>vector quantity</u> like velocity — it has <u>magnitude</u> and <u>direction</u>.

Acceleration — The Formula:

$$\text{acceleration} = \frac{\text{change in velocity}}{\text{time taken}}$$

Well, it's <u>just another formula</u>.
And it's got a <u>formula triangle</u> like all the others.

Mind you, there are <u>two tricky things</u> with this one:

First there's the 'v – u', which means working out the '<u>change in velocity</u>', as shown in the example below, rather than just putting a <u>simple value</u> for velocity or speed in.

Secondly there's the <u>unit</u> of acceleration, which is <u>m/s²</u>. <u>Not m/s</u>, which is <u>velocity</u>, but <u>m/s²</u>.

Here 'v' is the final velocity and 'u' is the initial velocity.

(v – u) could also be written Δv — it just means "change in v".

Example

> A skulking cat accelerates from 2 m/s to 6 m/s in 5.6 s. Find its acceleration.
>
> <u>ANSWER</u>: Using the formula triangle:
> a = (v – u) ÷ t = (6 – 2) ÷ 5.6 = 4 ÷ 5.6 = <u>0.71 m/s²</u>

Acceleration measures how quickly the velocity is changing

Acceleration is a <u>vector quantity</u> because it's got <u>magnitude</u> and <u>direction</u> — its units are <u>m/s²</u>, so if you do an acceleration calculation, don't forget to stick m/s² on the end of your answer.

Velocity-Time Graphs

Here's the distance-time graph's big brother — the <u>velocity-time</u> graph.

Velocity-Time Graphs

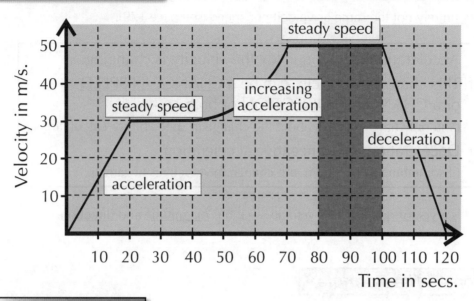

Very Important Notes:

1) <u>GRADIENT = ACCELERATION</u>.
2) <u>Flat sections</u> represent <u>steady speed</u>.
3) The <u>steeper</u> the graph, the <u>greater</u> the <u>acceleration</u> or <u>deceleration</u>.
4) <u>Uphill</u> sections (/) are <u>acceleration</u>.
5) <u>Downhill</u> sections (\\) are <u>deceleration</u>.
6) The <u>area</u> under any section of the graph (or all of it) is equal to the <u>distance travelled</u> in that <u>time interval</u>.
7) A <u>curve</u> means <u>changing acceleration</u>.

Calculating **Acceleration** and **Distance** From a **V-T** Graph

1) The <u>acceleration</u> represented by the <u>first section</u> of the graph is:

$$\underline{\text{Acceleration}} = \underline{\text{gradient}} = \frac{\text{vertical}}{\text{horizontal}} = \frac{30}{20} = \underline{1.5 \text{ m/s}^2}$$

This is the just the equation for acceleration (p.181).

2) The <u>distance travelled</u> in any time interval is equal to the <u>area</u> under the graph. For example, the distance travelled between t = 80 s and t = 100 s is equal to the <u>shaded area</u>, which is equal to $20 \times 50 = \underline{1000 \text{ m}}$.

3) You can only use this method for calculating the distance travelled for <u>uniform</u> (constant or steady) <u>acceleration</u>.

Don't get distance-time graphs and velocity-time graphs confused

So that's a velocity-time graph. You work out acceleration from the graph simply by applying the acceleration formula — change in velocity is the change on the vertical axis and time taken is the change on the horizontal axis. And once again, don't forget your units — m/s².

Weight, Mass and Gravity

Now for something a bit more attractive — the force of gravity.

Gravitational Force is the *Force of Attraction* Between *All Masses*

Gravity attracts all masses, but you only notice it when one of the masses is really really big, e.g. a planet. Anything near a planet or star is attracted to it very strongly.
This has two important effects:

1) On the surface of a planet, it makes all things accelerate (see p.181) towards the ground (all with the same acceleration, g, which is about 10 m/s² on Earth).

2) It gives everything a weight.

Weight and *Mass* are *Not The Same*

1) Mass is just the amount of 'stuff' in an object.
   For any given object this will have the same value anywhere in the universe.

2) Weight is caused by the pull of the gravitational force. In most questions the weight of an object is just the force of gravity pulling it towards the centre of the Earth.

3) An object has the same mass whether it's on Earth or on the Moon — but its weight will be different. A 1 kg mass will weigh less on the Moon (about 1.6 N) than it does on Earth (about 10 N), simply because the gravitational force pulling on it is less.

4) Weight is a force measured in newtons. It's measured using a spring balance or newton meter. Mass is not a force. It's measured in kilograms with a mass balance (an old-fashioned pair of balancing scales).

The *Very Important Formula* Relating *Mass*, *Weight* and *Gravity*

weight = mass × gravitational field strength

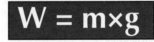

The acceleration due to gravity is always equal to the gravitational field strength, no matter what planet or moon you're on. That's why they can both be represented by 'g'.

1) Remember, weight and mass are not the same. Mass is in kg, weight is in newtons.
2) The letter "g" represents the strength of the gravity and its value is different for different planets. On Earth g ≈ 10 N/kg. On the Moon, where the gravity is weaker, g is only about 1.6 N/kg.
3) This formula is hideously easy to use:

> EXAMPLE: What is the weight, in newtons, of a 5 kg mass, both on Earth and on the Moon?
>
> ANSWER: "W = m × g". On Earth: W = 5 × 10 = 50 N   (The weight of the 5 kg mass is 50 N.)
>   On the Moon: W = 5 × 1.6 = 8 N   (The weight of the 5 kg mass is 8 N.)

See what I mean. Hideously easy — as long as you know what all the letters mean.

Warm-Up and Exam Questions

Here's another set of questions to test your knowledge.
Make sure you can answer them all before you go steaming on.

Warm-Up Questions

1) What does the gradient of a distance-time graph show?
2) What are the units of acceleration? and of mass? and of weight?
3) A car goes from 0 to 30 m/s in 6 seconds. Calculate its acceleration.
4) Name the force that makes things accelerate towards the surface of the Earth.

Exam Questions

1   A racing car is driven round a circular track of length 2400 m at a constant speed of 45 m/s.

   (a)   Explain why the car's velocity is not constant.

(1 mark)

   (b)   On one lap, the speed of the car increases from 45 m/s to 59 m/s over a period of
       5 seconds. Calculate its acceleration.

(2 marks)

2   The graph below shows the distance of a shuttle-bus from its start point plotted against time.

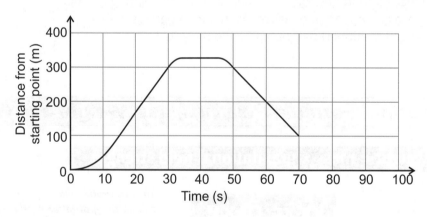

Use the graph to answer the following questions:

   (a)   Between 15 and 30 seconds:

      (i)   how far does the bus travel?

(1 mark)

      (ii)   how fast is the bus going?

(2 marks)

   (b)   For how long does the bus stop?

(1 mark)

   (c)   Describe the bus's speed and direction between 50 and 70 seconds.

(1 mark)

   (d)   Between 70 and 100 seconds, the bus slows, coming to a standstill at 100 s
       to finish up where it started. Show this on the graph.

(1 mark)

Exam Questions

3 The diagram below shows the velocity of a cyclist plotted against time.

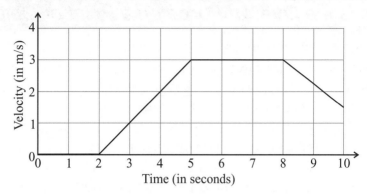

(a) Describe the motion of the cyclist between 5 and 8 seconds.

(1 mark)

(b) Describe what is happening to the cyclist's speed between 8 and 10 seconds.

(1 mark)

(c) Calculate how far the cyclist travelled between 2 and 5 seconds.

(1 mark)

4 A spring increases in length when masses are suspended from it, as shown. When a metal ball with a mass of 0.1 kg is suspended from the spring, the spring stretches by 3 cm.

If the experiment was repeated on Mars, the spring would only be stretched by 1.1 cm.

(a) Explain why the spring would stretch less on Mars, given that the value of the gravitational field strength is lower on Mars than on Earth.

(2 marks)

(b) Calculate an estimate for g on Mars, assuming that g on Earth is 10 m/s^2.

(3 marks)

5 A stone falls from the edge of a cliff. After falling for 1 second the stone has a downwards velocity of 10 m/s.

(a) Calculate the stone's acceleration during the first second it falls.

(2 marks)

(b) Assuming no air resistance, calculate the stone's velocity after three seconds of falling.

(2 marks)

(c) The stone has a mass of 0.12 kg. Calculate its weight.
Take the value of gravitational field strength to be 10 N/kg.

(2 marks)

(d) Describe the effect of doubling the stone's mass on its acceleration due to gravity.

(1 mark)

6 Which of the following masses exert a gravitational attraction on other masses — the Sun, the Earth, a human being, a feather, an atom? Explain your answer.

(1 mark)

Resultant Forces

You can work out how all the forces acting on an object add up together.

Resultant Force is the Overall Force on a Point or Object

The notion of resultant force is a bit tricky to get your head round:

1) In most real situations there are at least two forces acting on an object along any direction.

2) The overall effect of these forces will decide the motion of the object
— whether it will accelerate, decelerate or stay at a steady speed.

3) If you have a number of forces acting at a single point, you can replace them with a single force
(so long as the single force has the same effect on the motion as the original forces acting all together).

4) If the forces all act along the same line (they're all parallel and act in the same or the opposite direction), the overall effect is found by just adding or subtracting them.

5) The overall force you get is called the resultant force.

Example: Stationary Teapot — All Forces Balance

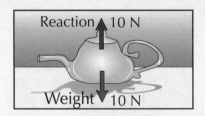

The length of the arrow shows the size of the force.

The direction of the arrow shows the direction of the force (didn't see that one coming, did you...).

If the arrows come in opposite pairs, and they're all the same size, then the forces are balanced.

1) The force of GRAVITY (or weight) is acting downwards.
2) This causes a REACTION FORCE (see p.188) from the surface pushing up on the object.
3) This is the only way it can be in BALANCE.
4) Without a reaction force, it would accelerate downwards due to the pull of gravity.
5) The resultant force on the teapot is zero: 10 N – 10 N = 0 N. *Remember that forces are always measured in newtons (N).*

A Resultant Force Means a Change in Velocity

1) If there is a resultant force acting on an object, then the object will change its state of rest or motion.

2) In other words it causes a change in the object's velocity.

You Can Find the Resultant Force Acting in a Straight Line

EXAMPLE: Benny is driving in his car. He applies a driving force of 1000 N, but has to overcome air resistance of 600 N.
What is the resultant force? Will the car's velocity change?

Driving Force: 1000 N   Air Resistance: 600 N   Resultant Force: 400 N

ANSWER: Say that the forces pointing to the left are pointing in the positive direction.
The resultant force = 1000 N – 600 N = 400 N to the left.
If there is a resultant force then there is always an acceleration, so Benny's velocity will change.

Forces and Acceleration

Around about the time of the Great Plague in the 1660s, a chap called <u>Isaac Newton</u> worked out his <u>Laws of Motion</u>. At first they might seem kind of obscure or irrelevant, but keep plugging away at it.

An Object Needs a **Force** to Start **Moving**

If the resultant force on a <u>stationary</u> object is <u>zero</u>, the object will <u>remain stationary</u>.

Things <u>don't just start moving</u> on their own, there has to be a <u>resultant force</u> (see p.186) to get them started.

No Resultant Force Means No Change in Velocity

If there is <u>no resultant force</u> on a <u>moving</u> object it'll just carry on moving at the <u>same velocity</u>.

1) When a train or car or bus or anything else is <u>moving</u> at a <u>constant velocity</u> then the <u>forces</u> on it must all be <u>balanced</u>.

2) Never let yourself entertain the <u>ridiculous idea</u> that things need a constant overall force to <u>keep</u> them moving — NO NO NO NO NO NO!

3) To keep going at a <u>steady speed</u>, there must be <u>zero resultant force</u> — and don't you forget it.

A Resultant Force Means Acceleration

If there is a <u>non-zero resultant force</u>, then the object will <u>accelerate</u> in the direction of the force.

1) A non-zero <u>resultant</u> force will always produce <u>acceleration</u> (or deceleration).

2) This "acceleration" can take <u>five</u> different forms: <u>Starting</u>, <u>stopping</u>, <u>speeding up</u>, <u>slowing down</u> and <u>changing direction</u>.

3) On a force diagram, the <u>arrows</u> will be <u>unequal</u>:

<u>Don't ever say</u>: "If something's moving there must be an overall resultant force acting on it". Not so. If there's an <u>overall</u> force it will always <u>accelerate</u>.

You get <u>steady</u> speed when there is <u>zero</u> resultant force.

I wonder how many times I need to say that same thing before you remember it?

Forces and Acceleration

More on forces and acceleration here. The big equation here is <u>F = ma</u> — it's a really important one. Remember that the F is always the <u>resultant force</u> — that's important too.

A *Non-Zero* Resultant Force Produces an *Acceleration*

Any <u>resultant force</u> will produce <u>acceleration</u>, and this is the <u>formula</u> for it:

$$F = ma \quad \text{or} \quad a = F/m$$

m = mass in kilograms (kg)   a = acceleration in metres per second squared (m/s²)
F is the <u>resultant force</u> in newtons (N)

> <u>EXAMPLE</u>: A car with a mass of 1750 kg has an engine which provides a driving force of 5200 N. At 70 mph the drag force acting on the car is 5150 N.
>
> Find its acceleration: a) when first setting off from rest   b) at 70 mph.

<u>ANSWER</u>: 1) First draw a force diagram for both cases (no need to show the vertical forces):

5200 N   0 mph   5200 N   5150 N   70 mph

2) Work out the resultant force and acceleration of the car in each case.

Resultant force = 5200 N   Resultant force = 5200 − 5150 = 50 N
a = F/m = 5200 ÷ 1750 = <u>3.0 m/s²</u>   a = F/m = 50 ÷ 1750 = <u>0.03 m/s²</u>

Reaction Forces are *Equal* and *Opposite*

> When <u>two objects interact</u>, the forces they exert on each other are <u>equal and opposite</u>.

These two forces are called an 'interaction pair'.

1) That means if you <u>push</u> something, say a shopping trolley, the trolley will <u>push back</u> against you, <u>just as hard</u>.

2) And as soon as you <u>stop</u> pushing, <u>so does the trolley</u>. Kinda clever really.

3) So far so good. The slightly tricky thing to get your head round is this — if the forces are always equal, <u>how does anything ever go anywhere</u>? The important thing to remember is that the two forces are acting on <u>different objects</u>.

Example — A Pair of Ice Skaters

When skater A pushes on skater B (the '<u>action</u>' force), she feels an equal and opposite force from skater B's hand (the '<u>reaction</u>' force).

Both skaters feel the <u>same sized force</u>, in <u>opposite directions</u>, and so accelerate away from each other.

Skater A will be <u>accelerated</u> more than skater B, though, because she has a smaller mass — remember <u>a = F/m</u>.

Skater A   Skater B
mass = 55 kg   mass = 65 kg

It's the same sort of thing when you go <u>swimming</u>. You <u>push</u> back against the <u>water</u> with your arms and legs, and the water pushes you forwards with an <u>equal-sized force</u> in the <u>opposite direction</u>.

Frictional Forces

Friction is found nearly everywhere and it acts to <u>slow down</u> and <u>stop</u> moving objects.
Sometimes friction is a pain, but at other times it's very helpful.

Friction is Always There To Slow Things Down

1) When an object is <u>moving</u> (or trying to move) friction acts in the direction that <u>opposes movement</u>.

2) The frictional force will <u>match</u> the size of the <u>force</u> trying to move it, <u>up to a point</u>
    — after this the friction will be <u>less</u> than the other force and the object will <u>move</u>.

3) <u>Friction</u> will act to make the moving object <u>slow down and stop</u>.

4) So to travel at a <u>steady speed</u>, things always need a <u>driving force</u> to overcome the friction.

5) Friction occurs in <u>three main ways</u>:

Friction Between Solid Surfaces Which Are Gripping

(static friction)

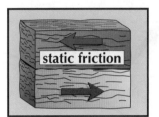

Friction Between Solid Surfaces Which are Sliding Past Each Other

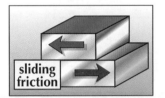

Resistance or "Drag" From Fluids (Liquids or Gases, e.g. Air)

Larger Area = More Drag

1) The larger the area of the object, the greater the drag.

2) So, to <u>reduce drag</u>, the area and <u>shape</u> should be <u>streamlined</u> and <u>reduced</u>, like <u>wedge-shaped</u>
    <u>sports cars</u>. <u>Roof boxes</u> on cars <u>spoil this shape</u> and so <u>slow them down</u>. Driving with the <u>windows</u>
    <u>open</u> also <u>increases drag</u>.

3) Something that's designed to reduce your speed, e.g. a <u>parachute</u>, often has a <u>large area</u> to give
    a <u>high drag</u> to slow you down (see next page).

4) For a given thrust, the <u>higher</u> the <u>drag</u>, the lower the <u>top speed</u>.

5) In a <u>fluid</u>, <u>friction</u> (drag) always <u>increases</u> as the <u>speed increases</u>.

Friction's annoying when it's slowing down your boat, car or lorry...

... but it can be useful too. As well as stopping parachutists ending up as nasty messes on the floor,
friction's good for <u>other stuff</u> — e.g. without it, you wouldn't be able to walk or run or skip or write.

Terminal Velocity

Frictional forces <u>increase</u> with speed — but only up to a certain point. Read on...

Moving Objects Can Reach a Terminal Velocity

1) When falling objects <u>first set off</u>, the force of gravity is <u>much more</u> than the <u>frictional force</u> slowing them down, so they accelerate.

2) As the <u>speed</u> increases, the friction <u>builds up</u>.

3) This gradually <u>reduces</u> the <u>acceleration</u> until eventually the <u>resistance force</u> is <u>equal</u> to the <u>accelerating force</u> and then it won't accelerate any more.

4) It will have reached its maximum speed or <u>terminal velocity</u> and will fall at a steady speed.

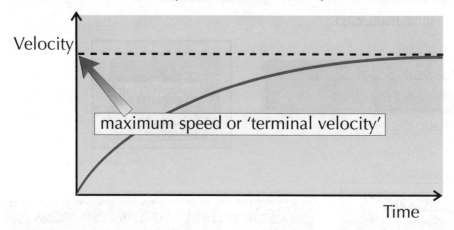

Terminal Velocity of Falling Objects Depends on Shape and Area

1) The <u>accelerating force</u> acting on <u>all</u> falling objects is <u>gravity</u> and it would make them all fall at the <u>same</u> rate, if it wasn't for <u>air resistance</u>.

2) This means that on the Moon, where there's <u>no air</u>, hamsters and feathers dropped simultaneously will hit the ground <u>together</u>.

3) However, on Earth, <u>air resistance</u> causes things to fall at <u>different</u> speeds, and the <u>terminal velocity</u> of any object is determined by its <u>drag</u> in <u>comparison</u> to its <u>weight</u>.

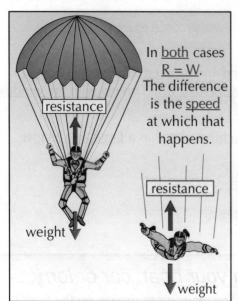

In <u>both</u> cases
R = W.
The difference is the <u>speed</u> at which that happens.

4) The frictional force depends on its <u>shape and area</u>.

5) The most important example is the human <u>skydiver</u>.

6) Without his parachute open he has quite a <u>small</u> area and a force of "<u>W = mg</u>" pulling him down.

7) He reaches a <u>terminal velocity</u> of about <u>120 mph</u>.

8) But with the parachute <u>open</u>, there's much more <u>air resistance</u> (at any given speed) and still only the same force "<u>W = mg</u>" pulling him down.

9) This means his <u>terminal velocity</u> comes right down to about <u>15 mph</u>, which is a <u>safe speed</u> to hit the ground at.

Warm-Up and Exam Questions

You're about halfway through this section and it's time for some more questions.

Warm-Up Questions

1) A rowing boat is being pulled to shore by two people with a force of 30 N each. A force of 10 N is resisting the movement in the opposite direction. What is the resultant force on the boat?

2) What is the resultant force on a body moving at constant velocity?

3) What happens to the acceleration of a body if the resultant force on it is doubled?

4) In which direction does friction act on a body — with or against the body's motion?

Exam Questions

1   Two parachutists, A and B, are members of the same club.

↑ 900 N

↓ 900 N

   (a)   The diagram shows the forces acting on parachutist A.

      (i)   What is the resultant force acting on parachutist A?

(1 mark)

      (ii)   Describe the velocity of parachutist A.

(1 mark)

   (b)   Parachutist B is in free fall.
      The total mass of parachutist B and her equipment is 70 kg.

      (i)   What will the force of air resistance on parachutist B be when she reaches terminal velocity? Explain your answer.

(3 marks)

      (ii)   Which parachutist, A or B, would have a higher terminal velocity? Explain your answer.

(3 marks)

   (c)   Explain why a parachutist slows down when they open their parachute.

(1 mark)

2   Stefan weighs 600 newtons. He is accelerating upwards in a lift at 2.5 m/s^2.

   (a)   The forces acting on Stefan are his weight and the upwards force exerted on him by the floor of the lift. Which force is greater? Explain your answer.

(2 marks)

   (b)   Calculate the size of the resultant force acting on Stefan.
      Take the value of gravitational field strength to be 10 N/kg.

(3 marks)

3   Damien's cricket bat has a mass of 1.2 kg. He uses it to hit a ball with a mass of 160 g forwards with a force of 500 N.

   (a)   State the force that the ball exerts on the bat.
      Explain your answer.

(2 marks)

   (b)   Which is greater — the acceleration of the bat or the ball? Explain your answer.

(2 marks)

Stopping Distances

If you brake suddenly, lots of factors affect how far you'll travel before you come to a complete stop...

Many Factors Affect Your Total Stopping Distance

1) If you need to stop in a given distance, then the faster a vehicle's going, the bigger braking force it'll need.

2) Likewise, for any given braking force, the faster you're going, the greater your stopping distance. But in real life it's not quite that simple — if your maximum braking force isn't enough, you'll go further before you stop.

3) The total stopping distance of a vehicle is the distance covered in the time between the driver first spotting a hazard and the vehicle coming to a complete stop.

4) The stopping distance is the sum of the thinking distance and the braking distance.

1) Thinking Distance

"The distance the vehicle travels during the driver's reaction time".

The reaction time is the time between the driver spotting a hazard and taking action.

It's affected by two main factors:

a) How fast you're going — Whatever your reaction time, the faster you're going, the further you'll go.

b) How dopey you are — This is affected by tiredness, drugs, alcohol and a careless blasé attitude.

Bad visibility and distractions can also be a major factor in accidents — lashing rain, messing about with the radio, bright oncoming lights, etc. might mean that a driver doesn't notice a hazard until they're quite close to it. It doesn't affect your thinking distance, but you start thinking about stopping nearer to the hazard, and so you're more likely to crash.

The figures below for typical stopping distances are from the Highway Code. It's frightening to see just how far it takes to stop when you're going at 70 mph.

2) Braking Distance

"The distance the car travels under the breaking force".

It's affected by five main factors:

a) How fast you're going — The faster you're going, the further it takes to stop.

b) How good your brakes are — All brakes must be checked and maintained regularly. Worn or faulty brakes will let you down catastrophically just when you need them the most, i.e. in an emergency.

c) The mass of your vehicle — With the same brakes, a heavily laden vehicle takes longer to stop.

d) How good the tyres are — Tyres should have a minimum tread depth of 1.6 mm in order to be able to get rid of the water in wet conditions. Leaves, diesel spills and muck on the road can greatly increase the braking distance, and cause the car to skid too.

e) How good the grip is — This depends on three things: 1) road surface, 2) weather conditions, 3) tyres.

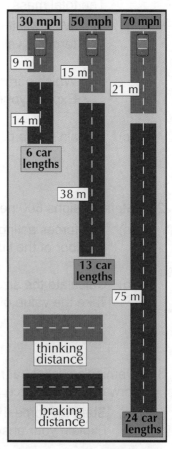

30 mph | 50 mph | 70 mph

9 m

15 m

21 m

14 m

6 car lengths

38 m

13 car lengths

75 m

thinking distance

braking distance

24 car lengths

Wet or icy roads are always much more slippy than dry roads, but often you only discover this when you try to brake hard. You don't have as much grip, so you travel further before stopping.

Momentum and Collisions

A <u>large</u> rugby player running very <u>fast</u> is going to be a lot harder to stop than a scrawny one out for a Sunday afternoon stroll — that's <u>momentum</u> for you.

Momentum = Mass × Velocity

1) Momentum (p) is a <u>property</u> of <u>moving objects</u>.
2) The <u>greater</u> the <u>mass</u> of an object and the <u>greater</u> its <u>velocity</u>, the <u>more momentum</u> the object has.
3) Momentum is a <u>vector</u> quantity — it has size <u>and</u> direction (like <u>velocity</u>, but not speed, see p.179).

Momentum (kg m/s) = Mass (kg) × Velocity (m/s)

*Momentum **Before** = Momentum **After***

In a <u>closed system</u>, the total momentum <u>before</u> an event (e.g. a collision) is the same as <u>after</u> the event. This is called <u>Conservation of Momentum</u>.

A <u>closed system</u> is just a fancy way of saying that no external forces act.

Example

Two skaters approach each other, collide and move off together as shown.
At what velocity do they move after the collision?

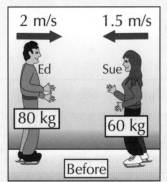

2 m/s	1.5 m/s	velocity (v) = ?
Ed	Sue	
80 kg	60 kg	(80+60) kg
	Before	After

1) Choose which direction is <u>positive</u>. I'll say "<u>positive</u>" means "<u>to the right</u>".
2) <u>Total momentum before</u> collision = momentum of Ed + momentum of Sue
   = {80 × 2} + {60 × (−1.5)}
   = <u>70 kg m/s</u>
3) <u>Total momentum after</u> collision = momentum of Ed and Sue together
   = <u>140 × v</u>
4) So 140v = 70, i.e. <u>v = 0.5 m/s to the right</u>.

Momentum's a pretty fundamental bit of Physics

Momentum is always <u>conserved</u> in collisions and explosions when there are no external forces acting.

Momentum and Collisions

Applying a <u>force</u> causes a <u>change</u> in <u>momentum</u>. For instance, if your annoying little brother is walking up to you and you push him (apply a force), he'll fall over (his momentum will change)...

Forces Cause Changes in Momentum

When a <u>force</u> acts on an object, it causes a <u>change</u> in momentum.

$$\text{Force (N)} = \frac{\text{Change in momentum (kg m/s)}}{\text{Time (s)}}$$

$$F = \frac{mv - mu}{t}$$

Here, 'v' is the final velocity, 'u' is the initial velocity and m is the mass.

(mv − mu) can also be written ΔM — it just means 'change in momentum'.

1) A <u>larger</u> force means a <u>faster</u> change of momentum (and so a greater <u>acceleration</u>, see page 181).

2) Likewise, if someone's momentum changes <u>very quickly</u> (like in a <u>car crash</u>), the <u>forces</u> on the body will be very <u>large</u>, and more likely to cause <u>injury</u>.

3) This is why cars are designed with protective features that slow people down over a <u>longer time</u> when they have a crash — the longer it takes for a change in <u>momentum</u>, the <u>smaller</u> the <u>force</u> (see page 203).

Example

A rock with mass <u>1 kg</u> is travelling through space at <u>15 m/s</u>.
A comet hits the rock, giving it a resultant force of <u>2500 N</u> for <u>0.7 seconds</u>.
Calculate   a) the rock's <u>initial momentum</u>, and
            b) the <u>change</u> in its momentum resulting from the impact.

ANSWER:
a) Momentum = mass × velocity = 1 × 15 = <u>15 kg m/s</u>
b) <u>Rearranging</u> the formula,
   Change of momentum = force × time = 2500 × 0.7 = <u>1750 kg m/s</u>.

Don't push your little brother over in the name of physics...

You see momentum in action all around you all the time. Momentum depends on <u>mass</u> and <u>velocity</u>, and a <u>force</u> can result in a <u>change in momentum</u>. Safety features on cars work by <u>slowing down</u> the change in momentum — there's more on this on page 203.

Work Done

Work (like a lot of things) means something slightly <u>different</u> in Physics than it does in everyday life.

Doing **Work** Involves **Transferring Energy**

When a <u>force</u> moves an <u>object</u> through a distance,
<u>ENERGY IS TRANSFERRED</u> and <u>WORK IS DONE</u>.

That statement sounds far more complicated than it needs to. Try this:

1) Whenever something <u>moves</u>, something else is providing some sort of '<u>effort</u>' to move it.

2) The thing putting the <u>effort</u> in needs a <u>supply</u> of <u>energy</u> (like <u>fuel</u> or <u>food</u> or <u>electricity</u> etc.).

3) It then does '<u>work</u>' by <u>moving</u> the object — and one way or another it <u>transfers</u> the energy it receives (as fuel) into <u>other forms</u>.

4) Whether this energy is transferred '<u>usefully</u>' (e.g. by <u>lifting a load</u>) or is '<u>wasted</u>' (e.g. lost as <u>heat</u> through <u>friction</u>), you can still say that '<u>work is done</u>'. Just like Batman and Bruce Wayne, '<u>work done</u>' and '<u>energy transferred</u>' are indeed '<u>one and the same</u>'. (And they're both given in <u>joules</u>.)

It's Just **Another Trivial Formula**:

work done = force × distance

Whether the force is <u>friction</u> or <u>weight</u> or <u>tension in a rope</u>, it's always the same. To find how much <u>energy</u> has been <u>transferred</u> (in joules), you just multiply the <u>force in N</u> by the <u>distance moved in</u> the direction of the force, measured in metres.

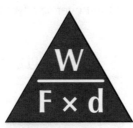

Example

Some kids drag an old tractor tyre 5 m over rough ground.
They pull with a total force of 340 N. Find the energy transferred.

<u>ANSWER</u>: Energy transferred is work done, so:
W = F × d = 340 × 5 = <u>1700 J</u>.

Remember "energy transferred" and "work done" are the same thing
By lifting something up you do work by transferring <u>chemical energy</u> to <u>gravitational potential energy</u> (p.201).

Power

Power is a concept that pops up in both <u>forces</u> and <u>electricity</u>. This is because, at its most fundamental level, power is just about the rate of <u>energy transfer</u> — and energy is transferred wherever you look.

Power is the "Rate of Doing Work" — i.e. How Much per Second

1) Power is <u>not</u> the same thing as <u>force</u>, nor <u>energy</u>.

2) A <u>powerful</u> machine is not necessarily one which can exert a strong <u>force</u> (though it usually ends up that way).

3) A <u>powerful</u> machine is one which transfers <u>a lot of energy in a short space of time</u>.

4) This is the <u>very easy formula</u> for power:

$$\text{Power} = \frac{\text{Work done (energy transferred)}}{\text{Time taken}}$$

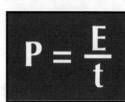

$$P = \frac{E}{t}$$

$$\frac{E}{P \times t}$$

Power is Measured in Watts (or J/s)

The proper unit of power is the <u>watt</u>. <u>One watt = 1 joule of energy transferred per second</u>. <u>Power</u> means "how much energy <u>per second</u>", so <u>watts</u> are the same as "<u>joules per second</u>" (J/s). Don't ever say "watts per second" — it's <u>nonsense</u>.

<u>EXAMPLE</u>: A motor transfers 4.8 kJ of useful energy in 2 minutes. Find its power output.

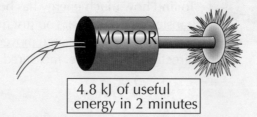

4.8 kJ of useful energy in 2 minutes

<u>ANSWER</u>: P = E ÷ t = 4800 ÷ 120 = 40 W (or 40 J/s)   *1 kJ = 1000J*

(Note that the kJ had to be turned into J, and the minutes into seconds.)

Watt is the unit of power?

Power is the amount of energy transferred per second, and it's measured in <u>watts</u>. The watt is named after James Watt, a Scottish inventor and engineer who did lots of work on steam engines in the 1700s.

Warm-Up and Exam Questions

There were lots of definitions and equations to get to grips with on the last five pages.
Try these questions to see what you can remember.

Warm-Up Questions

1) What is meant by 'thinking distance' as part of the total stopping distance of a car?
2) What must be added to thinking distance to find the total stopping distance of a car?
3) Give the equation for finding momentum.
4) What is meant by the conservation of momentum?
5) Why can 'work done' be measured in the same units as energy?
6) What is meant by power in terms of work done?

Exam Questions

1   The graph below shows how thinking distance and stopping distance vary with speed.

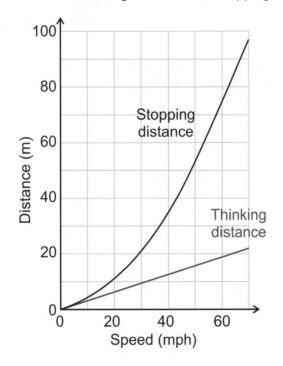

(a)   Use the graph to determine the following distances for a car travelling at 40 mph.

(i)   Thinking distance

(ii)   Total stopping distance

(iii)   Braking distance

(3 marks)

(b)   Which is greater at 50 miles per hour, thinking distance or braking distance?

(1 mark)

(c)   Is stopping distance proportional to speed?
Explain how this can be seen from the graph.

(1 mark)

If you need some help with part c) — see page 11.

Exam Questions

2   Two ice hockey players are skating towards the puck. Player A has a mass of 100 kg and is travelling right at 6 m/s. Player B has a mass of 80 kg and is travelling left at 9 m/s.

   (a) Calculate the momentum of:

      (i)   Player A

(2 marks)

      (ii) Player B

(2 marks)

   (b) The two players collide and become joined together.

      (i)   Calculate the speed of the two joined players just after the collision.

(3 marks)

      (ii) State the direction they move.

(1 mark)

3   A train is driven 700 m whilst accelerating at 1.05 m/s². The resultant force acting on the train is 42 000 N.

   (a) Calculate the work done by the resultant force.

(2 marks)

   (b) The train reaches a constant speed at which its kinetic energy is 29 400 000 J. It then decelerates with a braking force of 29 400 N. Calculate its braking distance.

(2 marks)

4   A fast-moving neutron collides with a uranium-235 atom and bounces off. The diagram shows the particles before and after the collision.

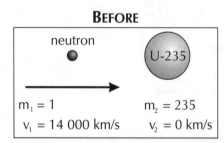

Find the velocity of the U-235 atom after the collision.

(3 marks)

5   A bus with a mass of 7000 kg is used to ferry passengers between terminals at an airport.

   (a) The engine of the bus has a power rating of 90 kW. Calculate the energy transferred by the engine in 5 seconds.

(2 marks)

   (b) Whilst travelling at 5 m/s, the bus hits a wall and comes to a complete stop in 0.5 s. Calculate the force acting on the bus during the collision.

(4 marks)

Kinetic Energy

Kinetic energy is the energy something has when it is moving. Think about it — if you bump into something heavy that's rushing downhill at top speed, it will do a lot more damage than if you bump into something light that's just sitting there.

Kinetic Energy is Energy of Movement

1) The kinetic energy (K.E.) of something is the energy it has when moving.

2) The kinetic energy of something depends on both its mass and speed.

3) The greater its mass and the faster it's going, the bigger its kinetic energy will be.

4) For example, a high-speed train, or a speedboat, will have lots of kinetic energy — but a scooter will only have a little bit.

5) Here's the formula for working out the kinetic energy of an object:

Sometimes kinetic energy is written as E_k.

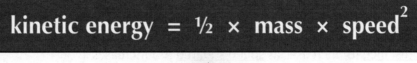

$$\text{kinetic energy} = \tfrac{1}{2} \times \text{mass} \times \text{speed}^2$$

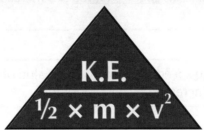

$$\frac{\text{K.E.}}{\tfrac{1}{2} \times m \times v^2}$$

Example

A car with a mass of 1450 kg is travelling at 28 m/s.
Calculate its kinetic energy.

28 m/s

Answer: It's pretty easy. You just plug the numbers into the formula — but watch the 'v^2'!

1450 kg

$$\text{K.E.} = \tfrac{1}{2}mv^2$$
$$= \tfrac{1}{2} \times 1450 \times 28^2$$
$$= \underline{568\ 400\ \text{J}}. \text{ (Joules because it's energy.)}$$

If you double the mass, the K.E. doubles. If you double the speed, though, the K.E. quadruples (increases by a factor of 4) — it's because of the 'v^2' in the formula. There's lots more about this coming up on the next page...

small mass, not fast
low kinetic energy

big fast
lorries Ltd

big mass, very fast
high kinetic energy

Kinetic Energy

Kinetic energy has a really big impact on stopping distances. It's mainly due to the $\underline{v^2}$ bit of the kinetic energy formula that you read about on the last page.

Stopping Distances *Increase Alarmingly* with *Extra Speed*

To stop a car, the kinetic energy, ½mv², has to be converted to heat energy at the brakes and tyres:

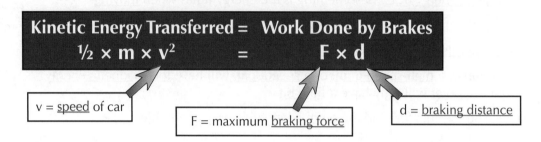

Kinetic Energy Transferred = Work Done by Brakes
½ × m × v² = F × d

v = speed of car

F = maximum braking force

d = braking distance

1) The braking distance (d) increases as speed squared (v²) increases — it's a squared relationship.

2) This means if you double the speed, you double the value of v, but the v² means that the K.E. is then increased by a factor of four.

3) Because 'F' is always the maximum possible braking force (which can't be increased), d must increase by a factor of four to make the equation balance.

Look back at page 192 for more on braking distances.

4) In other words, if you go twice as fast, the braking distance must increase by a factor of four to convert the extra K.E.

5) Increasing the speed by a factor of 3 increases the K.E. by a factor of $\underline{3^2}$ (= 9), so the braking distance becomes 9 times as long.

6) Doubling the mass of the object doubles the K.E. it has — which will double the braking distance. So a big heavy lorry will need more space to stop than a small car.

Kinetic energy — time to get a move on...

When meteors and space shuttles enter the atmosphere, they have loads and loads of kinetic energy. Friction with the air transfers kinetic energy to heat — so much heat that most meteors burn up completely and never hit us. Which is lucky, all things considered. Space shuttles have heat shields that lose heat quickly, so they can re-enter the atmosphere without burning up.

Gravitational Potential Energy

Gravitational potential energy is the energy an object has because of its height.
When something <u>falls</u>, this gravitational energy is converted into <u>kinetic energy</u>.

Gravitational Potential Energy is Energy Due to Height

gravitational potential energy = mass × g × height

1) <u>Gravitational potential energy</u> (measured in joules) is the energy
that an object has by virtue of (because of) its <u>vertical position</u>
in a <u>gravitational field</u>.

2) When an object is raised vertically, <u>work is done</u> against the
<u>force of gravity</u> (it takes effort to lift it up) and the object gains
gravitational potential energy.

3) On <u>Earth</u> the gravitational field strength (g) is approximately <u>10 N/kg</u>.

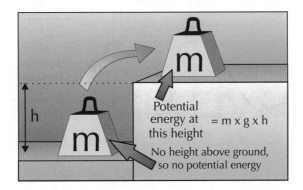

Example

A sheep of mass 47 kg is slowly raised through 6.3 m. Find the gain in potential energy.

<u>Answer:</u>   Just plug the numbers into the formula:
G.P.E. = m × g × h
= 47 × 10 × 6.3
= <u>2961 J</u> (Joules because it's <u>energy</u>.)

Falling Objects Convert G.P.E. into K.E.

1) When something <u>falls</u>, its <u>gravitational potential energy</u> is <u>converted</u> into
<u>kinetic energy</u> (<u>K.E.</u>). So the <u>further</u> it falls, the <u>faster</u> it goes.

2) In practice, some of the G.P.E. will be <u>dissipated</u> as <u>heat</u> due to <u>air resistance</u>,
but if you're told you can <u>ignore</u> air resistance you can just use this <u>simple</u>
and <u>really quite obvious formula</u>:

K.E. <u>gained</u> = G.P.E. <u>lost</u>

Energy and Roller Coasters

Roller coasters are a really good example of what happens when G.P.E. converts to K.E. and back again.

G.P.E. Lost is Equal to K.E. Gained (If you Ignore Air Resistance)

You saw on the last page that when an object falls, its gravitational potential energy (G.P.E.) is converted into kinetic energy (K.E.)...

1) For example, the roller coaster below will lose G.P.E. and gain K.E. as it falls between points A and C.

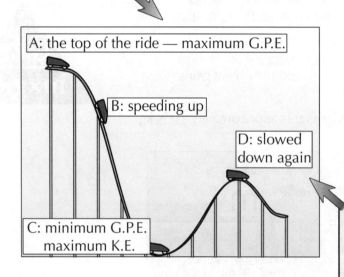

A: the top of the ride — maximum G.P.E.

B: speeding up

D: slowed down again

C: minimum G.P.E. maximum K.E.

2) If you ignore friction (between the tracks and the wheels) and air resistance, the amount of K.E. it gains will be the same as the amount of G.P.E. it loses.

3) Between C and D, it's gaining height, so some of that K.E. is converted back to G.P.E. again.

Example

The carriage in the diagram above has a mass of 500 kg and the vertical height difference between A and C is 20 m.

Weight is the force due to an object's mass and gravity. Weight = mass × acceleration due to gravity (about 10 m/s²).

a) Ignoring friction and air resistance, how much K.E. is gained by the carriage in moving from A to C?

b) The carriage was stationary at A. Calculate its speed at C.

ANSWER:

a) K.E. gained = G.P.E. lost = mass × g × height = 500 × 10 × 20 = 100 000 J

b) At C it has 100 000 J of K.E.

You know that K.E. = ½mv² (see page 199), so...

If you know how much K.E. something's gained you can calculate its speed.

$\frac{1}{2} \times m \times v^2 = 100\,000$

$v^2 = 100\,000 \div (\frac{1}{2} \times m) = 100\,000 \div (\frac{1}{2} \times 500) = 400$

$v = \sqrt{400} = 20$ m/s

Roller coasters make G.P.E. and K.E. fun

Roller coasters are constantly converting between gravitational potential and kinetic energy. In reality, energy will be lost due to friction, air resistance and even as sound. But you can usually ignore these. That's something to remember next time you're whizzing down a big dipper.

Car Design and Safety

Nowadays, cars usually come with lots of different safety features all designed to slow you down over a longer time in a crash. This page is all about how they work.

Cars are *Designed* to *Convert Kinetic Energy Safely* in a Crash

1) If a car crashes it will slow down very quickly — this means that a lot of kinetic energy is converted into other forms of energy in a short amount of time, which can be dangerous for the people inside.

2) In a crash, there'll be a big change in momentum (see page 194) over a very short time, so the people inside the car experience huge forces that could be fatal.

3) Cars are designed to convert the kinetic energy of the car and its passengers in a way that is safest for the car's occupants. They often do this by increasing the time over which momentum changes happen, which lessens the forces on the passengers.

Examples

CRUMPLE ZONES at the front and back of the car crumple up on impact.
- The car's kinetic energy is converted into other forms of energy by the car body as it changes shape.
- Crumple zones increase the impact time, decreasing the force produced by the change in momentum.

AIR BAGS also slow you down more gradually and prevent you from hitting hard surfaces inside the car.

SEAT BELTS stretch slightly, increasing the time taken for the wearer to stop. This reduces the forces acting in the chest. Some of the kinetic energy of the wearer is absorbed by the seat belt stretching.

SIDE IMPACT BARS are strong metal tubes fitted into car door panels. They help direct the kinetic energy of the crash away from the passengers to other areas of the car, such as the crumple zones.

Buckle up and read up — this could save your life

It's marvellous that cars nowadays come with so many safety features. Who knew physics could be so useful. Why don't you check out all the snazzy safety features next time you're in a car — such fun.

Car Design and Safety

Here's a bit more about the design of cars.
First up, a new type of <u>braking system</u> that <u>stores</u> energy rather than wasting it. How handy.

Brakes do *Work* Against the *Kinetic Energy* of the Car

1) When you <u>apply the brakes</u> to slow down a car, <u>work is done</u> (see p.195).

2) The brakes reduce the <u>kinetic energy</u> of the car by transferring it into <u>heat</u> (and sound) energy.

3) <u>ABS</u> (anti-lock <u>b</u>raking <u>s</u>ystem) <u>brakes</u> help drivers <u>keep control</u> of the car's <u>steering</u> when <u>braking hard</u>. They <u>automatically pump on and off</u> to stop the wheels locking and <u>prevent skidding</u>.

4) In <u>traditional</u> braking systems that would be the <u>end of the story</u>, but new <u>regenerative braking systems</u> used in some <u>electric</u> or <u>hybrid</u> cars <u>make use</u> of the energy, instead of converting it all into heat during braking.

Regenerative Braking Systems:

1) <u>Regenerative brakes</u> use the <u>system</u> that <u>drives</u> the vehicle to do the <u>majority of the braking</u>.

2) Rather than converting the kinetic energy of the vehicle into heat energy, the brakes put the vehicle's <u>motor into reverse</u>. With the motor running <u>backwards</u>, the wheels are <u>slowed</u>.

3) At the same time, the motor acts as an <u>electric generator</u>, converting kinetic energy into <u>electrical energy</u> that is stored as <u>chemical energy</u> in the vehicle's <u>battery</u>. This is the advantage of regenerative brakes — they <u>store</u> the energy of braking rather than <u>wasting</u> it. It's a nifty chain of energy transfer.

Cars Have Different *Power Ratings*

1) The <u>size</u> and <u>design</u> of car engines determine how <u>powerful</u> they are.

2) The <u>more powerful</u> an engine is, the more <u>energy</u> it transfers from its <u>fuel</u> every second, and so the <u>faster</u> its top speed can be.

3) E.g. the <u>power output</u> of a typical small car will be around 50 kW and a sports car will be about 100 kW (some are <u>much</u> higher).

Sports car power = 100 kW

Small car power = 50 kW

4) Cars are also designed to be <u>aerodynamic</u>. This means that they are shaped in such a way that <u>air flows</u> very easily and smoothly past them, so minimising their <u>air resistance</u>.

5) Cars reach their <u>top speed</u> when the resistive force <u>equals</u> the driving force provided by the engine (see p.189).

6) So, with <u>less air resistance</u> to overcome, the car can reach a <u>higher speed</u> before this happens. Aerodynamic cars therefore have <u>higher top speeds</u>.

The more powerful an engine, the faster it transfers energy from fuel

Now you know why cars are designed to be aerodynamic — it isn't just so they look better. Try bringing up these physics facts next time you're watching Top Gear — you'll be the envy of all your mates.

Forces and Elasticity

Forces aren't just important for cars and falling objects — you can <u>stretch things</u> with them as well.

Elastic Objects Store Energy as Elastic Potential Energy

1) When you apply a force to an object you may cause it to <u>stretch</u> and <u>change in shape</u>.

2) Any object that can <u>go back</u> to its <u>original shape</u> after the force has been removed is an <u>elastic object</u>.

3) <u>Work is done</u> to an elastic object to <u>change</u> its shape.
   This energy is not lost but is <u>stored</u> by the object as <u>elastic potential energy</u>.

4) The elastic potential energy is then <u>converted to kinetic energy</u> when the <u>force is removed</u> and the object returns to its original shape, e.g. when a spring or an elastic band bounces back.

Extension of an Elastic Object is Directly Proportional to Force...

If a spring is supported at the top and then a weight attached to the bottom, it <u>stretches</u>.

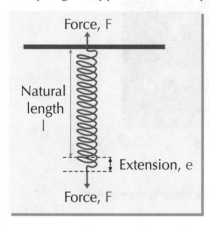

1) The <u>extension</u>, <u>e</u>, of a stretched spring (or other elastic object) is <u>directly proportional</u> to the load or <u>force</u> applied, <u>F</u>. The extension is measured in metres, and the force is measured in newtons.

2) This is the equation that links the <u>force applied</u> and the <u>extension</u> of the spring:

$$F = k \times e$$

3) k is the <u>spring constant</u>. Its value depends on the <u>material</u> that you are stretching and it's measured in newtons per metre (N/m).

...but this Stops Working when the Force is Great Enough

There's a <u>limit</u> to the amount of force you can apply to an object for the extension to keep on increasing <u>proportionally</u>.

1) The graph shows <u>force against extension</u> for an elastic object.

2) For small forces, force and extension are <u>proportional</u>. So the first part of the graph shows a straight-line relationship between force and extension.

3) There is a <u>maximum</u> force that the elastic object can take and still extend proportionally. This is known as the <u>limit of proportionality</u> and is shown on the graph at the point marked P.

4) If you increase the force <u>past</u> the limit of proportionality, the material will be <u>permanently stretched</u>. When the force is <u>removed</u>, the material will be <u>longer</u> than at the start.

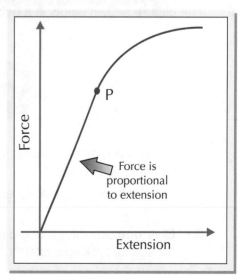

Warm-Up and Exam Questions

It's very nearly the end of this section. But don't shed a tear — try these questions instead.

Warm-Up Questions

1) Give the formula for calculating kinetic energy.
2) What is gravitational potential energy?
3) Why do the passengers in a car experience large forces during a crash?
4) Give two factors that will affect the top speed of a car.
5) What type of energy does an elastic object store when work is done to it?
6) What happens to a spring if you stretch it past its limit of proportionality?

Exam Questions

1   The images below show two different cars, car A and car B.

Car A

Car B

(a)   Car A has side impact bars.
Describe how side impact bars work during a crash to help protect passengers.

(2 marks)

(b)   Both cars have seat belts.
Explain why seat belts are made of a slightly stretchy material.

(3 marks)

(c)   Car A is more aerodynamic than car B.
Explain why aerodynamic cars tend to have higher top speeds.

(3 marks)

2   The picture shown below is a hybrid bus used for public transport in a city centre.

The bus has a regenerative braking system which is used to store energy
in the battery of the bus.

(a)   Explain how a regenerative braking system works.

(3 marks)

(b)   Give **one** advantage of regenerative braking systems over
traditional braking systems.

(1 mark)

Exam Questions

3   A car with a mass of 2750 kg is travelling at 12 m/s.

   (a) Calculate its kinetic energy.

(2 marks)

   (b) A van with a mass of 3120 kg is travelling at the same speed.
        Which has more kinetic energy?

(1 mark)

   (c) The car accelerates and reaches a constant speed at which its kinetic energy
        is 550 000 J. The car then brakes and comes to rest in 25 m.
        Calculate its braking force.

(2 marks)

4   A teacher is setting up an experiment.

   (a) He lifts a 2.5 kg mass from the floor onto a table that is 1.3 m tall.
        Calculate the gain in gravitational potential energy of the mass.
        (Use g = 10 N/kg.)

(2 marks)

   (b) The mass is accidentally knocked off the table and falls to the floor.
        Calculate the speed of the mass as it hits the floor.

(3 marks)

   (c) The teacher shows his students an experiment to show how
        a spring extends when masses are hung from it. When a force
        of 4 N is applied to the spring, the spring extends by 3.5 cm.

        Calculate the spring constant of the spring.
        Clearly show how you work out your answer.

(3 marks)

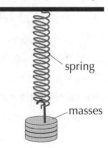
spring

masses

5   The Banshee X sports car can go from 100 mph to a complete stop in 10 s.

   (a) The manufacturer wants to add a luxury interior which will double the mass of the
        Banshee X. Describe what will happen to the braking distance of the car after the
        modification is made. Explain your answer.

(2 marks)

   (b) The Banshee X is fitted with ABS brakes.
        Explain why ABS brakes are safer to use than traditional brakes.

(2 marks)

   (c) When applied, the brakes reduce the kinetic energy of the car.
        What type of energy is the kinetic energy transferred into? Circle **two** answers.

   | Elastic Potential | Heat | Sound | Gravitational Potential |

(2 marks)

Revision Summary for Section 7

Well done — you've made it to the end of another section. There are loads of bits and bobs about forces, motion and fast cars which you have to learn. The best way to find out what you know is to get stuck in to these revision questions...

1) What's the difference between speed and velocity?

2) Sketch a typical distance-time graph and point out all the important parts of it.

3) Explain how to calculate speed from a distance-time graph.

4) What is acceleration?

5)* Write down the formula for acceleration. What's the acceleration of a soggy pea flicked from rest to a speed of 14 m/s in 0.4 seconds?

6) Sketch a typical velocity-time graph and point out all the important parts of it.

7) Explain how to find speed, distance and acceleration from a velocity-time graph.

8) Explain the difference between mass and weight.

9) Explain what is meant by a "resultant force".

10) If an object has zero resultant force on it, can it be moving? Can it be accelerating?

11)*Write down the formula relating resultant force and acceleration. A resultant force of 30 N pushes a trolley of mass 4 kg. What will be its acceleration?

12)*A skydiver has a mass of 75 kg. At 80 mph, the drag force on the skydiver is 650 N. Find the acceleration of the skydiver at 80 mph (take g = 10 N/kg).

13)*A yeti pushes a tree with a force of 120 N. What is the size of the reaction force that the Yeti feels pushing back at him?

14) What is "terminal velocity"?

15) If the total momentum of a system before a collision is zero, what is the total momentum of the system after the collision?

16)*Write down the formula for work done. A crazy dog drags a big branch 12 m over the next-door neighbour's front lawn, pulling with a force of 535 N. How much work was done?

17)*Find the kinetic energy of a 78 kg sheep moving at 23 m/s.

18)*A car of mass 1000 kg is travelling at a velocity of 2 m/s when a dazed and confused sheep runs out 5 m in front. If the driver immediately applies the maximum braking force of 395 N, can he avoid hitting it?

19)*A 4 kg cheese is taken 30 m up a hill before being rolled back down again. If g = 10 N/kg:
    a) how much gravitational potential energy does the cheese have at the top of the hill?
    b) how much gravitational potential energy does it have when it gets half way down?

20)*Calculate the kinetic energy of a 78 kg sheep just as she hits the floor after falling through 20 m.

21) Explain how crumple zones and air bags are useful in a crash.

22) Write down the equation that relates the force on a spring and its extension.

* Answers on page 278.

Static Electricity

Static electricity is all about charges which are <u>not</u> free to move, e.g. in insulating materials. This causes them to build up in one place and it often ends with a <u>spark</u> or a <u>shock</u> when they do finally move.

Build-up of **Static** is Caused by **Friction**

1) When certain <u>insulating</u> materials are <u>rubbed</u> together, negatively charged electrons will be <u>scraped off one</u> and <u>dumped</u> on the other.
2) This'll leave a <u>positive</u> static charge on one and a <u>negative</u> static charge on the other.
3) <u>Which way</u> the electrons are transferred <u>depends</u> on the <u>two materials</u> involved.
4) Electrically charged objects <u>attract</u> small objects placed near them.
5) The classic examples are <u>polythene</u> and <u>acetate</u> rods being rubbed with a <u>cloth duster</u>, as shown in the diagrams.

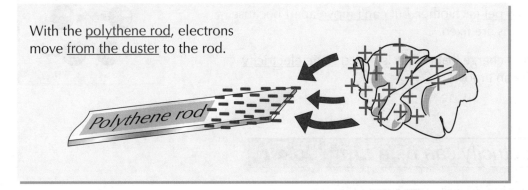

With the <u>polythene rod</u>, electrons move <u>from the duster</u> to the rod.

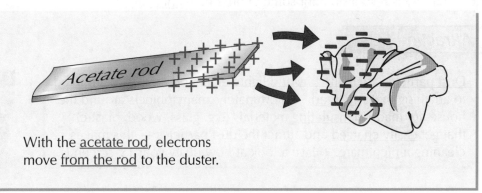

With the <u>acetate rod</u>, electrons move <u>from the rod</u> to the duster.

Only Electrons Move — *Never the Positive Charges*

1) Both +ve and −ve electrostatic charges are only ever produced by the movement of <u>electrons</u>.
2) The positive charges <u>definitely do not move</u>.
3) A positive static charge is always caused by electrons <u>moving</u> away elsewhere.
4) The material that <u>loses</u> the electrons loses some negative charge, and is <u>left with an equal positive charge</u> (see above). Don't forget!

Static electricity is caused by electrons being transferred

Static electricity's great fun. Come on, you must have tried it — rubbing a balloon against your jumper and trying to get it to stick to the ceiling. It really works... well, sometimes. Ahem.

Static Electricity

We come across static electricity all the time. From crackling clothes to bad hair days and other things in between, physics is here to explain it all.

Like Charges Repel, *Opposite* Charges Attract

Two things with <u>opposite</u> electric charges are <u>attracted</u> to each other. Two things with the <u>same</u> electric charge will <u>repel</u> each other.

1) When you rub two <u>insulating</u> materials together a whole load of <u>electrons</u> get dumped <u>together</u> on one of the insulators, which becomes <u>negatively charged</u>.

2) They try to <u>repel</u> each other, but <u>can't move</u> apart because their positions are fixed.

3) The patch of charge that results is called <u>static electricity</u> because it can't move.

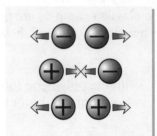

Static Electricity can be a *Little Joker*

Static electricity is responsible for some of life's little <u>annoyances</u>...

Attracting Dust

<u>Dust particles</u> are really tiny and lightweight and are easily <u>attracted</u> to anything that's <u>charged</u>. Unfortunately, many objects around the house are made of <u>insulating</u> materials (e.g. glass, wood, plastic) that get <u>easily charged</u> and attract the dust particles — this makes cleaning a <u>nightmare</u>. (Have a look at how dusty your TV screen is.)

Clinging Clothes and Crackles

When <u>synthetic clothes</u> are <u>dragged</u> over each other (like in a <u>tumble drier</u>) or over your <u>head</u>, electrons get scraped off, leaving <u>static charges</u> on both parts, and that leads to the inevitable — <u>attraction</u> (they stick together and cling to you) and little <u>sparks</u> or <u>shocks</u> as the charges <u>rearrange themselves</u>.

Bad Hair Days

Static builds up on your hair, giving each <u>strand</u> the same <u>charge</u> — so they <u>repel</u> each other.

Dangers of Static Electricity

Static Electricity can be *Dangerous*

1) A Lot of *Charge* can Build Up on *Clothes*

1) A large amount of <u>static charge</u> can build up on clothes made out of <u>synthetic materials</u> if they rub against other synthetic fabrics (see page 210).

2) Eventually, this <u>charge</u> can become <u>large enough</u> to make a <u>spark</u> — which is really bad news if it happens near any <u>inflammable gases</u> or <u>fuel fumes</u>... KABOOM!

2) *Grain Chutes*, *Paper Rollers* and the *Fuel Filling Nightmare*

1) As <u>fuel</u> flows out of a <u>filler pipe</u>, or <u>paper</u> drags over <u>rollers</u>, or <u>grain</u> shoots out of <u>pipes</u>, then <u>static can build up</u>.

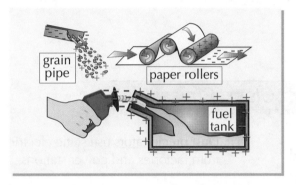

2) This can easily lead to a <u>spark</u> and might cause an explosion in <u>dusty</u> or <u>fumy</u> places — like when <u>filling up</u> a car with fuel at a <u>petrol station</u>.

3) All these problems can be solved by <u>earthing charged objects</u>...

Objects Can be *Earthed* or *Insulated* to *Prevent Sparks*

1) Dangerous <u>sparks</u> can be prevented by connecting a charged object to the <u>ground</u> using a <u>conductor</u> (e.g. a copper wire) — this is called <u>earthing</u> and it provides an <u>easy route</u> for the static charges to travel into the ground. This means <u>no charge</u> can build up to give you a shock or make a spark.

2) Static charges are a <u>big problem</u> in places where sparks could ignite <u>inflammable gases</u>, or where there are high concentrations of <u>oxygen</u> (e.g. in a <u>hospital</u> operating theatre).

3) <u>Fuel tankers</u> must be <u>earthed</u> to prevent any sparks that might cause the fuel to <u>explode</u> — <u>refuelling aircraft</u> are <u>bonded</u> to their fuel tankers using an <u>earthing cable</u> to prevent sparks.

4) <u>Anti-static sprays</u> and liquids work by making the surface of a charged object <u>conductive</u> — this provides an <u>easy path</u> for the charges to <u>move away</u> and not cause a problem.

5) Anti-static <u>cloths</u> are conductive, so they can carry charge away from objects they're used to <u>wipe</u>.

6) <u>Insulating mats</u> and shoes with <u>insulating soles</u> prevent static electricity from <u>moving</u> through them, so they stop you from getting a <u>shock</u>.

Static electricity — it's really shocking stuff...

<u>Lightning</u> is an extreme case of a static electric spark. It always chooses the <u>easiest path</u> between the sky and the ground — that's why it's never a good idea to fly a kite in a thunderstorm...

Uses of Static Electricity

Static electricity isn't always a nuisance. It's actually got loads of applications in <u>medicine</u> and <u>industry</u>.

Paint Sprayers *give an Even Coating*

1) Bikes and cars are painted using <u>electrostatic paint sprayers</u>.

2) The spray gun is <u>charged</u>, which charges up the small drops of paint. Each paint drop <u>repels</u> all the others, since they've all got the <u>same charge</u>, so you get a very <u>fine spray</u>.

3) The object to be painted is given an <u>opposite charge</u> to the gun and <u>attracts</u> the fine spray of paint.

4) This method gives an <u>even coat</u> and hardly any paint is <u>wasted</u>.

5) Also, even the parts of the bicycle or car that are pointing <u>away</u> from the spray gun <u>still receive paint</u> too — there are no paint <u>shadows</u>.

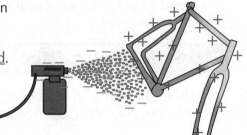

Dust Precipitators *Clean Emissions*

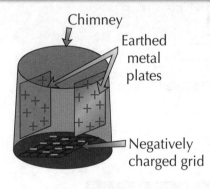

Chimney

Earthed metal plates

Negatively charged grid

1) <u>Dust precipitators</u> use static electricity to clean up <u>emissions</u> from factories and power stations.

2) <u>Dust particles</u> become <u>negatively charged</u> as they pass through a <u>charged wire grid</u> in the chimney.

3) The negatively charged dust particles then stick to <u>earthed metal plates</u> and eventually fall to the bottom of the chimney where they can be removed.

Defibrillators *give Electric Shocks*

1) An electric shock from a <u>defibrillator</u> can <u>restart</u> a stopped heart.

2) The defibrillator consists of two <u>paddles</u> connected to a power supply which are placed <u>firmly</u> on the patient's chest.

3) The defibrillator operator holds <u>insulated handles</u> — so <u>only the patient</u> gets a shock.

4) The <u>charge</u> passes through the <u>paddles</u> to the patient to make the <u>heart contract</u>.

Electric shocks aren't always bad

Yeah, I know they're not the most comfortable of things but they do have loads of uses in the real world. Who'd have thought that an electric shock could actually save someone's life...

Warm-Up and Exam Questions

By this point you'll probably have worked out that static electricity isn't the most exciting of topics. Don't worry — there are just these few questions before you get on to much more interesting stuff.

Warm-Up Questions

1) Is a polythene rod an insulator or a conductor?
2) What are the two types of electric charge?
3) Do similar charges attract or repel one another?
4) Give an example of an object that can be painted using an electrostatic paint sprayer.

Exam Questions

1   Jane hangs an uncharged balloon from a thread. She brings
    a negatively charged polythene rod towards the balloon.
    The diagram below shows how the positive and negative
    charges in the balloon rearrange themselves when she does this.

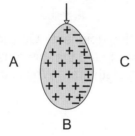

(a)   In which of the positions labelled A, B and C on the diagram
      did Jane hold the polythene rod? Explain your answer.

(2 marks)

(b)   Jane brings the rod closer to the balloon.
      Explain why the balloon swings towards it.

(1 mark)

2   A positive static charge builds up on a cloth when it is used to wipe a surface.

(a)   Describe the movement of charged particles that gives the cloth its charge.

(1 mark)

(b)   The cloth has a relative charge of +23.
      Circle the correct answer below to show the charge on the surface.

| +23 | +46 | -46 | -23 |

(1 mark)

(c)   The cloth is an insulator so charges can't easily flow through it.
      Give **one** example of a material that charges can flow easily through.

(1 mark)

Exam Questions

3   The office where Jyoti works has a nylon carpet.

(a)   Jyoti often wears rubber-soled shoes to work.

(i)   What effect of static electricity might happen if she touches a metal chair leg?

(1 mark)

(ii)   Explain why this would happen.

(2 marks)

(b)   Jyoti uses a device called a Network Interface Card (NIC) to allow her laptop computer to connect to the office wi-fi network. The NIC can be harmed by static electricity.

(i)   Why does the instruction manual for the NIC recommend that she touches a metal chair leg before touching the NIC?

(1 mark)

(ii)   Would touching a wooden chair leg work in the same way? Explain your answer.

(1 mark)

4   Mike is decorating a picture frame using gold leaf. Gold leaf is made of gold that has been beaten into very thin sheets. Mike has found that his brush picks up the pieces of gold leaf better if he rubs the bristles against a cloth first. Why is this?

(2 marks)

5   Defibrillators like the one shown, are used on patients who have suffered a cardiac arrest.

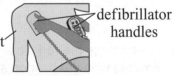

(a)   What are defibrillators used for?

(1 mark)

(b)   What sort of material are the handles of the defibrillator made of — insulating or conducting? Why is this?

(1 mark)

(c)   Name another device that uses static electricity.

(1 mark)

6   Static electricity can be dangerous.

(a)   Use your knowledge of static electricity to suggest why using an unearthed filler pipe to fill up a plane with fuel could be dangerous.

(2 marks)

(b)   Explain how fitting the fuel tanker with an earthing cable could reduce this danger.

(2 marks)

Current

Isn't <u>electricity</u> great. Mind you it's pretty bad news if the <u>words</u> don't mean anything to you...

Some Important *Terms*

1) <u>Current</u> is the <u>flow</u> of electric charge round the circuit. Current will <u>only flow</u> through a component if there is a <u>potential difference</u> across that component. Unit: ampere, A.

2) <u>Potential difference</u> is the <u>driving force</u> that pushes the current round. It's also called the voltage. Unit: volt, V.

3) <u>Resistance</u> is anything in the circuit which <u>slows the flow down</u>. Unit: ohm, Ω.

4) There's a <u>balance</u> — the <u>voltage</u> is trying to <u>push</u> the current round the circuit, and the <u>resistance</u> is <u>opposing</u> it — the <u>relative sizes</u> of the voltage and resistance decide <u>how big</u> the current will be:

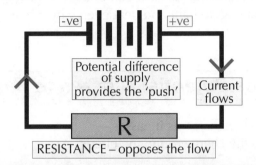

> If you <u>increase the voltage</u> — then <u>more current</u> will flow.
> If you <u>increase the resistance</u> — then <u>less current</u> will flow
> (or <u>more voltage</u> will be needed to keep the <u>same current</u> flowing).

Total Charge *Through a Circuit Depends on* Current *and* Time

1) <u>Current</u> is the <u>rate of flow</u> of <u>charge</u>. When <u>current</u> (I) flows past a point in a circuit for a length of <u>time</u> (t) then the <u>charge</u> (Q) that has passed is given by this formula:

$$\text{Current} = \frac{\text{Charge}}{\text{Time}} \qquad I = \frac{Q}{t}$$

2) <u>Current</u> is measured in <u>amperes</u> (A), <u>charge</u> is measured in <u>coulombs</u> (C), <u>time</u> is measured in <u>seconds</u> (s).

3) In the <u>metal wires</u> of a circuit, this charge is carried by <u>electrons</u>. Metals are <u>good conductors</u> as they have <u>free electrons</u> which are able to move.

4) <u>More charge</u> passes around the circuit when a <u>bigger current</u> flows.

> <u>EXAMPLE:</u> A battery charger passes a current of 2.5 A through a cell over a period of 4 hours. How much charge does the charger transfer to the cell altogether?
>
> <u>ANSWER:</u> Q = I × t = 2.5 × (4 × 60 × 60) = 36 000 C (36 kC).

More resistance means less current

Just remember that increasing the resistance slows down the flow. It's a bit like being prepared for an exam — if you don't resist revision and learn your stuff, the answers will come flowing out...

Potential Difference

Potential Difference (P.D.) is the Work Done Per Unit Charge

1) The potential difference (or <u>voltage</u>) is the <u>work done</u> (the energy transferred, measured in joules, J) <u>per coulomb of charge</u> that passes between <u>two points</u> in an electrical circuit.

2) It's given by this formula:

$$\text{P.D.} = \frac{\text{Work done}}{\text{Charge}}$$

3) So, the potential difference across an electrical component is the <u>amount of energy</u> that is transferred by that electrical component (e.g. to light and heat energy by a bulb) <u>per unit of charge</u>.

A Voltmeter Measures Potential Difference Between Two Points

1) A <u>battery</u> transfers energy <u>to</u> the charge as it passes — that's the "<u>push</u>" that moves the charge round the circuit.

2) <u>Components</u> transfer energy <u>away from</u> the charge as it passes — e.g. to use as <u>light</u> in a lamp or <u>sound</u> in a buzzer.

3) The voltage of a battery shows <u>how much</u> work the battery will do to charge that passes <u>through it</u> (how big a "<u>push</u>" it gives it).

4) A <u>voltmeter</u> is used to measure the <u>potential difference</u> between <u>two points</u>.

5) A voltmeter must be placed in <u>parallel</u> (see p.224) with a component so it can <u>compare</u> the energy the charge has <u>before</u> and <u>after</u> passing through the component (as in the diagram below).

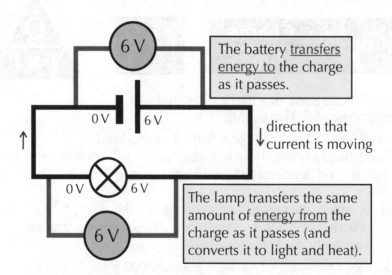

The battery <u>transfers energy to</u> the charge as it passes.

direction that current is moving

The lamp transfers the same amount of <u>energy from</u> the charge as it passes (and converts it to light and heat).

I think it's about time you took charge...

A very slightly interesting fact — voltage is named after Count Alessandro Volta, an Italian physicist. It's not a classic factoid, but you never know when it might come in handy later in life...

Circuits — The Basics

Formulas are mighty pretty and all, but you might have to design some <u>electrical circuits</u> as well one day. For that you're going to need <u>circuit symbols</u>...

Each Component Has a Circuit Symbol

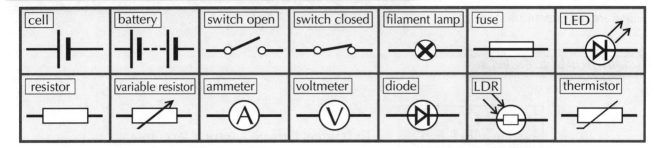

The Standard Test Circuit

This is the circuit you use if you want to know the <u>resistance of a component</u>.
You find the resistance by measuring the <u>current through</u> and the <u>potential difference across</u> the component. It is absolutely the most <u>bog standard</u> circuit you'll come across.

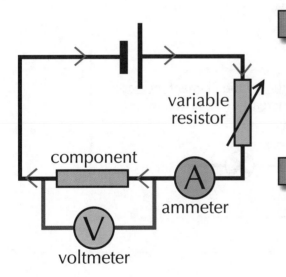

The Ammeter

1) Measures the <u>current</u> (in <u>amps</u>) flowing through the component.

2) Must be placed <u>in series</u> (see p.222).

3) Can be put <u>anywhere</u> in series in the <u>main circuit</u>, but <u>never</u> in parallel like the voltmeter.

The Voltmeter

1) Measures the <u>potential difference</u> (in <u>volts</u>) across the component.

2) Must be placed <u>in parallel</u> (see p.224) around the <u>component</u> under test — <u>NOT</u> around the variable resistor or the battery.

Five Important Points

1) This <u>very basic</u> circuit is used for testing <u>components</u>, and for getting <u>V-I graphs</u> from them (see next page).

2) The <u>component</u>, the <u>ammeter</u> and the <u>variable resistor</u> are all in <u>series</u>, which means they can be put in <u>any order</u> in the main circuit. The <u>voltmeter</u>, on the other hand, can only be placed <u>in parallel</u> around the <u>component under test</u>, as shown. Anywhere else is a definite <u>no-no</u>.

3) As you <u>vary</u> the <u>variable resistor</u> it alters the <u>current</u> flowing through the circuit.

4) This allows you to take several <u>pairs of readings</u> from the <u>ammeter</u> and <u>voltmeter</u>.

5) You can then <u>plot</u> these values for <u>current</u> and <u>voltage</u> on a <u>V-I graph</u> and find the <u>resistance</u>.

V-I Graphs

With your current and your potential difference measured, you can now make some graphs...

Three Hideously Important Potential Difference-Current Graphs

V-I graphs show how the <u>current</u> varies as you <u>change</u> the <u>potential difference</u> (P.D.).
Here are three examples:

Different Resistors

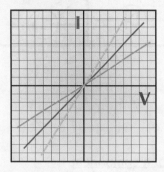

The current through a <u>resistor</u> (at constant temperature) is <u>directly proportional to P.D.</u>
<u>Different resistors</u> have different <u>resistances</u>, hence the different <u>slopes</u>.

Filament Lamp

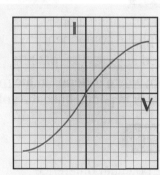

As the <u>temperature</u> of the filament <u>increases</u>, the <u>resistance increases</u> (see next page), hence the <u>curve</u>.

Diode

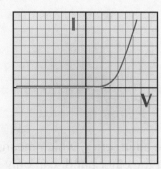

Current will only flow through a diode in <u>one direction</u>, as shown. The diode has very <u>high resistance</u> in the opposite direction.

V-I graphs have distinctive shapes

Understanding the shapes of these graphs can be a little tricky at first but have no fear — there's more on why these graphs are the shapes they are on pages 219 and 220.

Resistance and V = I × R

Prepare yourself to meet the <u>most important</u> equation in electrics, bar none.
But first up, a bit more about resistance...

Resistance *Increases* with *Temperature*

1) When an electrical charge flows through a resistor, some of the electrical
   energy is <u>transferred to heat energy</u> and the resistor gets <u>hot</u>.

2) This heat energy causes the <u>ions</u> in the conductor to <u>vibrate more</u>.

3) This makes it <u>more difficult</u> for the charge-carrying electrons to get through the
   resistor — the <u>current can't flow</u> as easily and the <u>resistance increases</u>.

4) A <u>filament lamp</u> contains a piece of wire with a really <u>high</u> resistance. When current passes
   through it, its <u>temperature increases</u> so much that it <u>glows</u> — which is the <u>light</u> you see.

5) For most resistors there is a <u>limit</u> to the amount of current that can flow.

6) More current means an <u>increase</u> in <u>temperature</u>, which means an
   <u>increase</u> in <u>resistance</u>, which means the <u>current decreases</u> again.

7) This is why the graph for the filament lamp <u>levels off</u> at high currents (see previous page).

Resistance, *Potential Difference* and *Current*: V = I × R

Potential Difference = Current × Resistance

For the <u>straight-line graphs</u> on the previous page, the resistance of the
component is <u>steady</u> and is equal to the <u>inverse</u> of the <u>gradient</u> of the line, or
"<u>1/gradient</u>". In other words, the <u>steeper</u> the graph the <u>lower</u> the resistance.

If the graph <u>curves</u>, it means the resistance is <u>changing</u>. In that case R can be found for any point
by taking the <u>pair of values</u> (V, I) from the graph and sticking them in the formula <u>R = V/I</u>. Easy.

Example

> Voltmeter V reads 6 V and resistor R is 4 Ω.
> What is the current through Ammeter A?
>
> <u>ANSWER</u>: Use the formula triangle for V = I × R.
> We need to find I, so the version we need is I = V/R.
> The answer is then: I = 6 ÷ 4 = <u>1.5 A</u>.

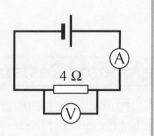

Be upstanding for the all important formula: V = I × R

Well OK, maybe not. It is a really handy formula though, and knowing about it makes understanding
the shapes of V–I graphs easier (see page 218). And that's got to be a good thing, in anyone's book.

Circuit Devices

You might consider yourself a bit of an <u>expert</u> in circuit components — you're enlightened about bulbs, you're switched on to switches... Well here are some more — they're a <u>bit trickier</u>.

Current Only Flows in **One Direction** through a **Diode**

1) A diode is a special device made from <u>semiconductor</u> material such as <u>silicon</u>.

2) It is used to <u>regulate</u> the <u>potential difference</u> in circuits.

3) It lets current flow freely through it in <u>one direction</u>, but <u>not</u> in the other (i.e. there's a very high resistance in the <u>reverse</u> direction).

4) This turns out to be really useful in various <u>electronic circuits</u>.

Light-Emitting Diodes are Very Useful

1) A <u>light-emitting diode</u> (LED) emits light when a current flows through it in the <u>forward direction</u>.

2) LEDs are being used more and more as lighting, as they use a much <u>smaller current</u> than other forms of lighting.

3) LEDs indicate the presence of current in a circuit. They're often used in appliances (e.g. TVs) to show that they are <u>switched on</u>.

4) They're also used for the numbers on <u>digital clocks</u>, in <u>traffic lights</u> and in <u>remote controls</u>.

A Light-Dependent Resistor or "LDR"

1) An LDR is a resistor that is <u>dependent</u> on the <u>intensity</u> of <u>light</u>. Simple really.

2) In <u>bright light</u>, the resistance <u>falls</u>.

3) In <u>darkness</u>, the resistance is <u>highest</u>.

4) They have lots of applications including <u>automatic night lights</u>, outdoor lighting and <u>burglar detectors</u>.

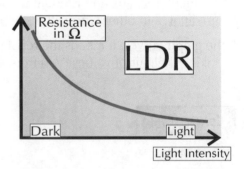

Thermistor Resistance Decreases as Temperature Increases

1) A <u>thermistor</u> is a <u>temperature dependent</u> resistor.

2) In <u>hot</u> conditions, the resistance <u>drops</u>.

3) In <u>cool</u> conditions, the resistance goes <u>up</u>.

4) Thermistors make useful <u>temperature detectors</u>, e.g. <u>car engine</u> temperature sensors and electronic <u>thermostats</u>.

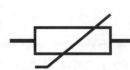

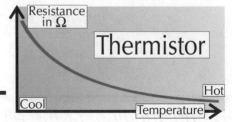

Warm-Up and Exam Questions

Phew — circuits aren't the easiest thing in the world, are they? Make sure you've understood the last few pages by trying these questions. If you get stuck, just go back and re-read the relevant page.

Warm-Up Questions

1) What are the units of resistance?
2) Write down the formula that links potential difference, work done and charge.
3) Draw the symbol for a light-emitting diode (LED).
4) Give one use of a light-dependent resistor (LDR).
5) What happens to the resistance of a thermistor as temperature increases?

Exam Questions

1   Shown below is a circuit diagram for a standard test circuit.

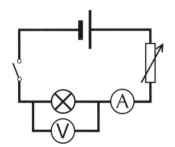

(a) When the switch is closed, the ammeter reads 0.3 A and the voltmeter reads 1.5 V.

  (i) Calculate the resistance of the filament lamp.

(2 marks)

  (ii) The switch is closed for 35 seconds. Calculate the total charge that flows through the filament lamp.

(2 marks)

(b) The variable resistor is used to increase the resistance in the circuit. Describe how this will affect the current flowing through the circuit.

(1 mark)

(c) The resistance of a filament lamp changes with temperature.

  (i) On the graph to the right, sketch the potential difference-current graph for a filament lamp.

(1 mark)

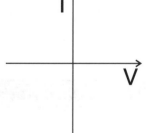

  (ii) Explain why the resistance of the filament lamp increases as the temperature of the filament increases.

(3 marks)

2   The graph below shows current against potential difference (P.D.) for a diode.

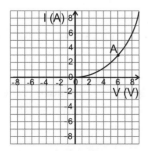

(a) Explain why the graph shows zero current for negative P.D.s.

(1 mark)

(b) Calculate the resistance of the diode at the point marked A.

(2 marks)

(c) Light emitting diodes (LEDs) are often used for lighting. Give **one** advantage of using LEDs instead of filament lamps.

(1 mark)

Series Circuits

You can tell the difference between series and parallel circuits <u>just by looking at them</u>.

Series Circuits — Everything in a Line

In <u>series circuits</u>, the different components are connected <u>in a line</u>, <u>end to end</u>, between the +ve and –ve of the power supply (except for <u>voltmeters</u>, which are always connected <u>in parallel</u>, but they don't count).

1) Potential Difference is **Shared**:

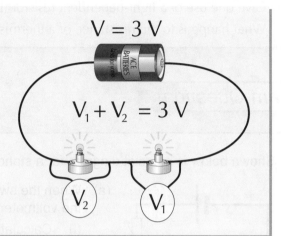

1) In series circuits, the <u>total potential difference</u> (P.D.) of the <u>supply</u> is <u>shared</u> between the various <u>components</u>. So the <u>P.D.s</u> round a series circuit always <u>add up</u> to equal the P.D. across the <u>battery</u>:

$$V = V_1 + V_2$$

2) This is because the total <u>work done on</u> the charge by the <u>battery</u> must equal the total <u>work done by</u> the charge on the <u>components</u>.

$$V = 3 \, V$$

$$V_1 + V_2 = 3 \, V$$

2) Current is the **Same** Everywhere:

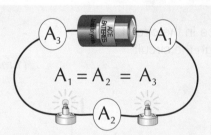

$$A_1 = A_2 = A_3$$

1) In series circuits the <u>same current</u> flows through <u>all parts</u> of the circuit: $\boxed{A_1 = A_2 = A_3}$

2) The <u>size</u> of the current is determined by the <u>total P.D.</u> of the cells and the <u>total resistance</u> of the circuit: i.e. $I = V/R$. This means <u>all</u> the <u>components</u> get the same <u>current</u>.

3) Resistance **Adds Up**:

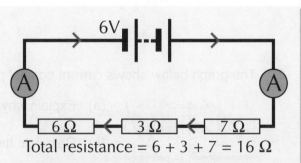

1) In series circuits, the <u>total resistance</u> is just the <u>sum</u> of the individual resistances:

$$R = R_1 + R_2 + R_3$$

6V

6 Ω   3 Ω   7 Ω

Total resistance = 6 + 3 + 7 = 16 Ω

2) The resistance of <u>two</u> (or more) resistors in <u>series</u> is <u>bigger</u> than the resistance of just one of the resistors on its own because the <u>battery</u> has to <u>push charge</u> through <u>all</u> of them.

3) The <u>bigger</u> the resistance of a component, the bigger its <u>share</u> of the <u>total P.D.</u> because more <u>work is done</u> by the charge when moving through a <u>large</u> resistance, than through a <u>small</u> one.

4) If the resistance of <u>one</u> component <u>changes</u> (e.g. if it's a variable resistor, light-dependent resistor or thermistor) then the <u>potential difference</u> across <u>all</u> the components will change too.

Series Circuits

And here are a few more rules that series circuits follow...

Cell Voltages **Add Up**

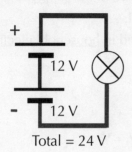

Total = 24 V

1) If you connect <u>several cells in series</u>, <u>all the same way</u> (+ to −) you get a <u>bigger total voltage</u> — because each charge in the circuit passes though all the cells and gets a 'push' from each cell in turn.

2) So <u>two 1.5 V</u> cells <u>in series</u> would supply <u>3 V in total</u>.

Cell voltages <u>don't</u> add up like that for cells connected <u>in parallel</u>. Each charge only goes through <u>one cell</u>.

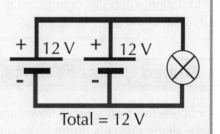

Total = 12 V

Cell Current **Doesn't Add Up**

1) Adding cells in <u>series doesn't</u> <u>increase the current</u> in a circuit.

2) The <u>maximum current</u> in the circuit will just be the <u>same</u> as if you had <u>one cell</u> in the circuit.

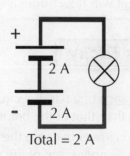

Total = 2 A

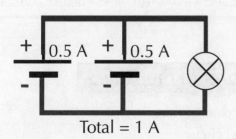

Total = 1 A

1) Cells connected in <u>parallel increase</u> <u>the total current</u> in the circuit.

2) However, the current through <u>each cell</u> is <u>less</u> than in the rest of the circuit because they <u>join together</u> to make the total current.

Parallel Circuits

Parallel circuits are much more <u>sensible</u> than series circuits. First up, the reason why...

Parallel Circuits — *Independence* and *Isolation*

1) In <u>parallel circuits</u>, each component is <u>separately</u> connected to the +ve and –ve of the <u>supply</u>.

2) If you remove or disconnect <u>one</u> of them, it will <u>hardly affect</u> the others at all.

3) This is <u>obviously</u> how <u>most</u> things must be connected, for example in <u>cars</u> and in <u>household electrics</u>. You have to be able to switch everything on and off <u>separately</u>.

1) P.D. is the *Same* Across *All* Components

1) In parallel circuits <u>all</u> components get the <u>full source P.D.</u>, so the voltage is the <u>same</u> across all components:

$$V_1 = V_2 = V_3$$

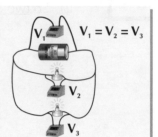

$$V_1 = V_2 = V_3$$

2) This means that <u>identical bulbs</u> connected in parallel will all be at the <u>same brightness</u>.

2) Current is *Shared* Between Branches

1) In parallel circuits the <u>total current</u> flowing around the circuit is equal to the <u>total</u> of all the currents through the <u>separate components</u>.

$$A = A_1 + A_2 + ...$$

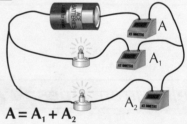

2) In a parallel circuit, there are <u>junctions</u> where the current either <u>splits</u> or <u>rejoins</u>. The total current going <u>into</u> a junction has to equal the total current <u>leaving</u>.

$$A = A_1 + A_2$$

3) If two <u>identical components</u> are connected in parallel then the <u>same current</u> will flow through each component.

3) Resistance Is *Tricky*

1) The <u>total resistance</u> of a parallel circuit is <u>tricky to work out</u>, but it's always <u>less</u> than that of the branch with the <u>smallest</u> resistance.

2) The resistance is lower because the charge has <u>more than one</u> branch to take — only <u>some</u> of the charge will flow along each branch.

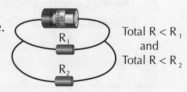

Total R < R_1
and
Total R < R_2

3) A circuit with two resistors in <u>parallel</u> will have a <u>lower</u> resistance than a circuit with either of the resistors <u>by themselves</u> — which means the <u>parallel</u> circuit will have a <u>higher current</u>.

Voltmeters and *Ammeters* Are *Exceptions* to the Rule:

1) Ammeters and voltmeters are <u>exceptions</u> to the series and parallel rules.

2) Ammeters are <u>always</u> connected in <u>series</u> even in a parallel circuit.

3) Voltmeters are <u>always</u> connected in <u>parallel with a component</u> even in a series circuit.

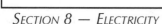

Mains Electricity and Oscilloscopes

Electric current is the <u>movement of charge carriers</u>. To transfer energy, it <u>doesn't matter which way</u> the charge carriers are going. That's why an <u>alternating current</u> works. Read on to find out more...

Mains Supply is *AC*, Battery Supply is *DC*

1) The UK mains supply is approximately <u>230 volts</u>.

2) It is an <u>AC supply</u> (alternating current), which means the current is <u>constantly</u> changing direction.

3) The frequency of the AC mains supply is <u>50 cycles per second</u> or <u>50 Hz</u> (hertz).

4) By contrast, cells and batteries supply <u>direct current</u> (DC).
   This just means that the current always keeps flowing in the <u>same direction</u>.

Electricity Supplies Can Be Shown on an *Oscilloscope* Screen

1) A <u>cathode ray oscilloscope</u> (CRO) is basically a <u>voltmeter</u>.

2) If you plug an <u>AC supply</u> into an oscilloscope, you get a '<u>trace</u>' on the screen that shows how the voltage of the supply changes with <u>time</u>. The trace goes up and down in a <u>regular pattern</u> — some of the time it's positive and some of the time it's negative.

3) If you plug in a <u>DC supply</u>, the trace you get is just a <u>straight line</u>.

4) The <u>vertical height</u> of the AC trace at any point shows the <u>input voltage</u> at that point.
   By measuring the height of the trace you can find the potential difference of the AC supply.

5) For DC it's a <u>lot simpler</u> — the voltage is just the distance from the <u>straight line trace</u> to the centre line.

There's more on oscilloscopes on the next page.

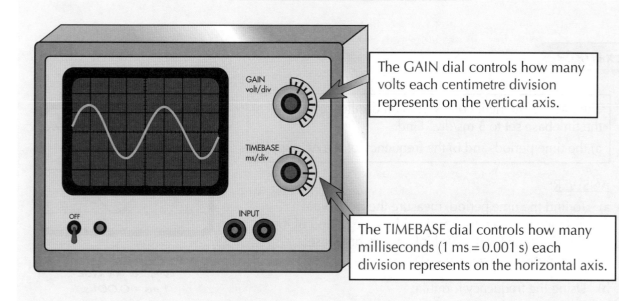

The GAIN dial controls how many volts each centimetre division represents on the vertical axis.

The TIMEBASE dial controls how many milliseconds (1 ms = 0.001 s) each division represents on the horizontal axis.

Height of AC trace = input voltage

AC current takes a while to get used to. Constantly switching the current between positive and negative may seem a bit pointless now but it'll all become clearer later on in the book...

Oscilloscopes

It's Pretty Easy to **Read** an **Oscilloscope Trace**

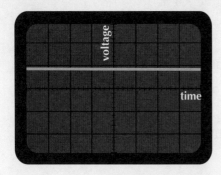

A <u>DC source</u> is always at the <u>same voltage</u>, so you get a <u>straight line</u>.

An <u>AC source</u> gives a <u>regularly repeating wave</u>. From that, you can work out the <u>period</u> and the <u>frequency</u> of the supply.

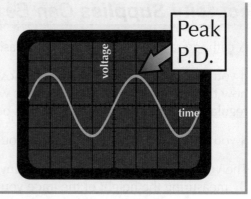

Peak P.D.

You work out the frequency using:

$$\text{Frequency (Hz)} = \frac{1}{\text{Time period (s)}}$$

Example

The trace to the right comes from an oscilloscope with the timebase set to 5 ms/div. Find:

a) the time period, and b) the frequency of the AC supply.

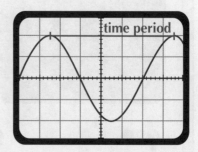

time period

Time period = the time to complete one cycle.
1 ms = 0.001 s.

<u>ANSWER</u>:

a) To find the time period, measure the horizontal distance between two peaks. The time period of the signal is 6 divisions. Multiply by the timebase:
Time period = 5 ms × 6 = <u>0.03 s</u>

b) Using the frequency formula:
Frequency = 1 ÷ 0.03 = <u>33 Hz</u>

Frequency is 1 over the time period

Because mains power is AC, its current can be increased or decreased using a device called a transformer (see pages 237-238). The lower the current in power transmission lines, the less energy is wasted as heat.

Warm-Up and Exam Questions

Those last few pages had lots more stuff on circuits and electricity.
Try these out to see what you can remember...

Warm-Up Questions

1) Give two things that determine the size of the current in a series circuit.
2) Are circuits in cars connected in series or parallel?
3) What is the frequency of UK mains electricity supply?
4) Do batteries supply alternating current (AC) or direct current (DC)?
5) What is the time period of a wave with a frequency of 100 Hz?

Exam Questions

1 The diagram below shows a series circuit.

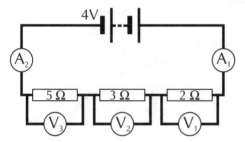

(a) Calculate the total resistance in the circuit.

(2 marks)

(b) The current through A_1 is 0.4 A. What is the current through A_2? Explain your answer.

(2 marks)

(c) V_1 reads 0.8 V and V_2 reads 1.2 V. Calculate the reading on V_3.

(2 marks)

2 The diagram on the right shows a trace on a CRO.

(a) Is the trace displaying the output from the mains or a battery? Explain your answer.

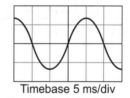

Timebase 5 ms/div

(1 mark)

(b) What is the time period of the wave?

(1 mark)

(c) What is the frequency of the wave?

(2 marks)

(d) What will happen to the CRO trace if the voltage of the supply is reduced?

(1 mark)

3 A parallel circuit is connected as shown.
Calculate the readings on:

(a) Voltmeter V_1.

(1 mark)

(b) Ammeter A_1.

(2 marks)

(c) Ammeter A_2.

(2 marks)

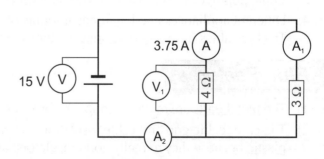

Electricity in the Home

It's important to know how to correctly <u>wire a plug</u> — plugs that aren't wired correctly are <u>dangerous</u>.

Hazards in the Home

Identifying <u>electrical hazards</u> in the home should be mostly <u>common sense</u>, but it helps if you already know some of the likely hazards, so here are 9 examples:

1) <u>Long cables</u>.
2) <u>Frayed cables</u>.
3) <u>Cables</u> in contact with something <u>hot</u> or <u>wet</u>.
4) <u>Water near sockets</u>.
5) <u>Shoving</u> things into sockets.

6) <u>Damaged plugs</u>.
7) <u>Too many</u> plugs into one socket.
8) Lighting sockets <u>without bulbs in</u>.
9) Appliances without their <u>covers</u> on.

Most Cables Have Three Separate Wires

1) Most electrical appliances are connected to the mains supply by <u>three-core</u> cables. This means that they have <u>three wires</u> inside them, each with a <u>core of copper</u> and a <u>coloured plastic coating</u>.

2) The brown <u>LIVE WIRE</u> in a mains supply alternates between a <u>HIGH +VE AND –VE VOLTAGE</u>.

3) The blue <u>NEUTRAL WIRE</u> is always at <u>0V</u>. Electricity normally flows in and out through the live and neutral wires only.

4) The green and yellow <u>EARTH WIRE</u> is for protecting the wiring, and for safety — it works together with a fuse to prevent fire and shocks. It is attached to the metal casing of the appliance and <u>carries the electricity to earth</u> (and away from you) should something go wrong and the live or neutral wires touch the metal case.

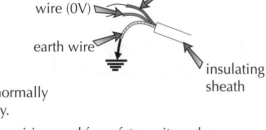

live wire (alternating between +ve and –ve high voltage)

neutral wire (0V)

earth wire

insulating sheath

Three-Pin Plugs and Cables Have Safety Features

Get the Wiring Right

1) The <u>right coloured wire</u> is connected to each pin, and <u>firmly screwed</u> in.
2) <u>No bare wires</u> showing inside the plug.
3) <u>Cable grip</u> tightly fastened over the cable <u>outer layer</u>.
4) Different appliances need <u>different</u> amounts of electrical energy. <u>Thicker</u> cables have <u>less resistance</u>, so they carry <u>more current</u>.

Rubber or plastic case

Earth Wire Green/Yellow

Fuse

Neutral Wire Blue

Live Wire Brown

Cable grip

Brass Pins

Plug Features

1) The <u>metal parts</u> are made of copper or brass because these are <u>very good conductors</u>.
2) The case, cable grip and cable insulation are made of <u>rubber</u> or <u>plastic</u> because they're really good <u>insulators</u>, and <u>flexible</u> too.
3) This all keeps the electricity flowing <u>where it should</u>.

Fuses and Earthing

Fuses are really useful devices so it'd be worth knowing a bit about them.

Earthing and Fuses Prevent Electrical Overloads

The earth wire and fuse (or circuit breaker) are included in electrical appliances for safety and work together like this:

1) If a fault develops in which the live wire somehow touches the metal case, then because the case is earthed, too great a current flows in through the live wire, through the case and out down the earth wire.

2) This surge in current melts the fuse (or trips the circuit breaker in the live wire) when the amount of current is greater than the fuse rating. This cuts off the live supply and breaks the circuit.

3) This isolates the whole appliance, making it impossible to get an electric shock from the case. It also prevents the risk of fire caused by the heating effect of a large current.

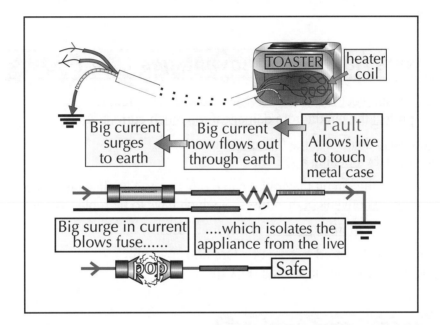

4) As well as people, fuses and earthing are there to protect the circuits and wiring in your appliances from getting fried if there is a current surge.

5) Fuses should be rated as near as possible but just higher than the normal operating current.

6) The larger the current, the thicker the cable you need to carry it. That's why the fuse rating needed for cables usually increases with cable thickness.

Fuses — they're everywhere

Safety precautions on modern appliances mean it's pretty difficult to get electrocuted by them. But that's only so long as they are in good condition and you're not doing something really stupid. Watch out for frayed wires, don't overload plugs, and for goodness sake don't use a knife to get toast out of a toaster when it is switched on. Turn over to the next page to find out more about safety precautions that you might find in your home.

Fuses and Earthing

Earthing is another way in which we can prevent appliances giving us an electric shock.

Insulating Materials Make Appliances "Double Insulated"

1) All appliances with metal cases are usually "earthed" to reduce the danger of electric shock.

2) "Earthing" just means the case must be attached to an earth wire.

3) An earthed conductor can never become live.

4) If the appliance has a plastic casing and no metal parts showing then it's said to be double insulated.

5) Anything with double insulation like that doesn't need an earth wire — just a live and neutral.

6) Cables that only carry the live and neutral wires are known as two-core cables.

Circuit Breakers Have Some Advantages Over Fuses

1) Circuit breakers are an electrical safety device used in some circuits. Like fuses, they protect the circuit from damage if too much current flows.

2) When circuit breakers detect a surge in current in a circuit, they break the circuit by opening a switch.

3) A circuit breaker (and the circuit they're in) can easily be reset by flicking a switch on the device. This makes them more convenient than fuses — which have to be replaced once they've melted.

4) They are, however, a lot more expensive to buy than fuses.

5) One type of circuit breaker used instead of a fuse and an earth wire is a Residual Current Circuit Breaker (RCCB):

Residual Current Circuit Breakers

1) Normally exactly the same current flows through the live and neutral wires. If somebody touches the live wire, a small but deadly current will flow through them to the earth. This means the neutral wire carries less current than the live wire. The RCCB detects this difference in current and quickly cuts off the power by opening a switch.

2) They also operate much faster than fuses — they break the circuit as soon as there is a current surge — no time is wasted waiting for the current to melt a fuse. This makes them safer.

3) RCCBs even work for small current changes that might not be large enough to melt a fuse. Since even small current changes could be fatal, this means RCCBs are more effective at protecting against electrocution.

Energy in Circuits

Electricity is just another form of <u>energy</u> — which means that it is always <u>conserved</u>.

*Energy is **Transferred** from Cells and Other **Sources***

Anything which <u>supplies electricity</u> is also supplying <u>energy</u>.

So cells, batteries, generators, etc. all <u>transfer energy</u> to components in the circuit:

| <u>Motion</u>: motors | <u>Light</u>: light bulbs | <u>Heat</u>: Hair dryers/kettles | <u>Sound</u>: speakers |

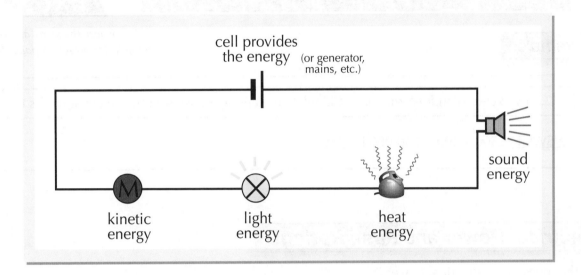

cell provides
the energy (or generator, mains, etc.)

kinetic energy

light energy

heat energy

sound energy

*All **Resistors** Produce **Heat** When a **Current** Flows Through Them*

1) Whenever a <u>current</u> flows through anything with <u>electrical resistance</u> (which is pretty much everything) then <u>electrical energy</u> is converted into <u>heat energy</u>.

2) The <u>more current</u> that flows, the more heat is produced.

3) A <u>bigger voltage</u> means more heating because it pushes more current through.

4) <u>Filament bulbs</u> work by passing a current through a very <u>thin wire</u>, heating it up so much that it glows. Rather obviously, they waste a lot of energy as <u>heat</u>.

*If an Appliance is **Efficient** it **Wastes Less Energy***

All this energy wasted as heat can get a little <u>depressing</u> — but there is a solution.

1) When you buy electrical appliances you can choose to buy ones that are more <u>energy efficient</u>.

2) These appliances transfer more of their <u>total electrical energy output to useful energy</u>.

3) For example, less energy is wasted as heat in power-saving lamps such as <u>compact fluorescent lamps</u> (CFLs) and <u>light-emitting diodes</u> (p.220) than in ordinary filament bulbs.

4) Unfortunately, they do <u>cost more to buy</u>, but over time the money you <u>save</u> on your electricity bills pays you back for the initial investment.

Power in Circuits

Power ratings tell you how much energy a device transfers per second.

Power Ratings of Appliances

1) The total energy transferred by an appliance depends on
how long the appliance is on and its power rating.

2) The power of an appliance is the energy that it uses per second.

Energy Transferred = Power rating × time

*This is just the same as the
formula for power on page 196.*

Example

A 2.5 kW kettle is on for 5 minutes. Calculate the energy transferred by the kettle in this time.

ANSWER: 2500 × 300 = 750 000 J = 750 kJ. (5 minutes = 300 s).

Electrical Power and Fuse Ratings

1) The formula for electrical power is:

POWER = CURRENT × POTENTIAL DIFFERENCE

$$P = I \times V$$

2) Most electrical goods show their power rating and voltage rating. To work out the size of
the fuse needed, you need to work out the current that the item will normally use.

Example

A hair dryer is rated at 230 V, 1 kW. Find the fuse needed.

ANSWER: I = P/V = 1000/230 = 4.3 A. Normally, the fuse should be rated just a
little higher than the normal current, so a 5 amp fuse is ideal for this one.

Use fuses with a rating just above the usual current

In the UK, you can usually get fuses rated at 3 A, 5 A or 13 A, and that's about it.
You should bear that in mind when you're working out fuse ratings.
If you find you need a 10.73 A fuse — tough. You'll have to use a 13A one.

Charge and Energy Change

You can think about electrical circuits in terms of energy transfer — the charge carriers take charge around the circuit, and when they go through an electrical component energy is transferred to make the component work. Read on for more...

Potential Difference is the Energy Transferred per Charge Passed

1) When an electrical charge (Q) goes through a change in potential difference (V), then energy (E) is transferred.

2) Energy is supplied to the charge at the power source to 'raise' it through a potential.

3) The charge gives up this energy when it 'falls' through any potential drop in components elsewhere in the circuit.

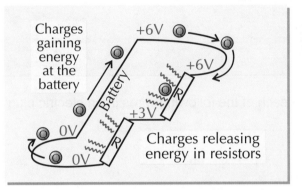

Charges gaining energy at the battery

+6V

+6V

+3V

0V

0V

Charges releasing energy in resistors

4) The formula is simple:

Energy transformed = Charge × Potential difference

$$\frac{E}{Q \times V}$$

5) The bigger the change in P.D. (or voltage), the more energy is transferred for a given amount of charge passing through the circuit.

6) That means that a battery with a bigger voltage will supply more energy to the circuit for every coulomb of charge which flows round it, because the charge is raised up "higher" at the start — and as the diagram shows, more energy will be dissipated in the circuit too.

Example

The motor in an electric toothbrush is attached to a 3 V battery.
If a current of 0.8 A flows through the motor for 3 minutes:

    a) Calculate the total charge passed.

    b) Calculate the energy transformed by the motor.

    c) Explain why the kinetic energy output of the motor will be less than your answer to b).

ANSWER: a) Use the formula (p.215) Q = I × t = 0.8 × (3 × 60) = <u>144 C</u>

          b) Use E = Q × V = 144 × 3 = <u>432 J</u>

          c) The motor won't be 100% efficient.
              Some of the energy will be transformed into <u>sound and heat</u>.

Warm-Up and Exam Questions

You're over half way through this section now. Check you can do the straightforward stuff with this warm-up, then have a go at the exam questions below.

Warm-Up Questions

1) Which of the live, neutral or earth wires is always at 0 volts?
2) Why is the case of a plug usually made out of plastic?
3) Appliances with double insulation don't need which type of wire?
4) What energy transformation occurs when electric current flows through a resistor?
5) What is the equation linking Q, V and E?

Exam Questions

1   (a)   What colour(s) are each of the following wires in an electric plug?
     (i)   live
     (ii)   neutral
     (iii) earth

(3 marks)

   (b)   Which wire alternates between a high positive and negative voltage?

(1 mark)

   (c)   What type of safety device contains a wire that is designed to melt when the current passing through it goes above a certain value?

(1 mark)

2   *In this question you will be assessed on the quality of your English, the organisation of your ideas and your use of appropriate specialist vocabulary.*

A domestic appliance has a plug containing live, neutral and earth wires and a fuse. The appliance has a metal case.

Describe how the earth wire and fuse work together to protect the appliance and to prevent the user getting an electric shock if there is a fault.

(6 marks)

3   A current of 0.5 A passes through a torch bulb. The torch is powered by a 3 V battery.
   (a)   What is the power of the torch?

(2 marks)

   (b)   If the torch is on for half an hour, how much charge has passed through the battery?

(2 marks)

   (c)   How much electrical energy does the bulb transfer in half an hour?

(2 marks)

Electromagnetic Induction

It's difficult to imagine a world <u>without electricity</u> — it would be hard to bake cakes at night, for a start.

Moving a Magnet into a Coil of Wire Induces a Voltage

1) You can create a <u>voltage</u>, and maybe a <u>current</u>, in a conductor by <u>moving a magnet</u> in or near a <u>coil of wire</u>. This is called <u>electromagnetic induction</u>.

2) As you <u>move</u> the magnet, the <u>magnetic field</u> through the <u>coil</u> changes — this <u>change</u> in the magnetic field <u>induces</u> (creates) a <u>voltage</u> across the <u>ends</u> of the coil.

3) If the ends of the wire are <u>connected</u> to make a closed circuit then a <u>current</u> will flow in the wire.

4) The <u>direction</u> of the voltage depends on which way you move the magnet:

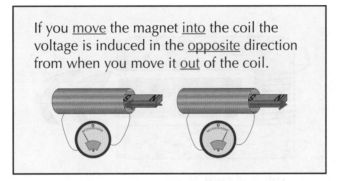

If you <u>move</u> the magnet <u>into</u> the coil the voltage is induced in the <u>opposite</u> direction from when you move it <u>out</u> of the coil.

If you <u>reverse</u> the magnet's North-South polarity — so that the opposite <u>pole</u> points into the coil, the voltage is induced in the <u>opposite</u> direction.

Four Factors Affect the Size of the Induced Voltage

If you want a <u>bigger</u> peak voltage (and current) you could do one or more of these <u>four things</u>...

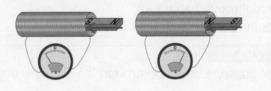

1) **Add** an **IRON CORE** inside the coil
2) **Increase** the **STRENGTH** of the **MAGNETIC FIELD**
3) **Increase** the **SPEED** of **ROTATION**
4) **Increase** the number of **TURNS** on the **COIL**

To <u>reduce the voltage</u>, you would <u>reduce</u> one of the <u>factors</u> or take the iron core out.

There are four factors that affect the induced voltage

It's important that you understand this page before you go on to read the rest of the pages in this section — moving a magnet into a coil of wire induces a voltage. Got it. Good.

AC Generators

Well, this page is actually more interesting than it looks (not that that would be very difficult). Haven't you always wondered how we make alternating current...

AC Generators — Just Turn the **Magnet** and There's a Current

1) In a generator, a <u>magnet</u> (or an <u>electromagnet</u>) <u>rotates</u> in a coil of wire.

2) As the magnet <u>turns</u>, the <u>magnetic field</u> through the <u>coil</u> changes.

3) This <u>change</u> in the magnetic field induces a <u>voltage</u>, which makes a <u>current</u> flow in the coil.

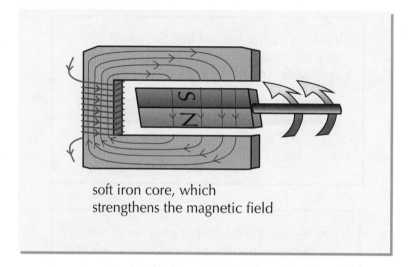

soft iron core, which
strengthens the magnetic field

4) When the magnet is turned through half a turn,
the <u>direction</u> of the <u>magnetic field</u> through the coil <u>reverses</u>.

5) When this happens, the <u>voltage reverses</u>,
so the <u>current</u> flows in the <u>opposite direction</u> around the coil of wire.

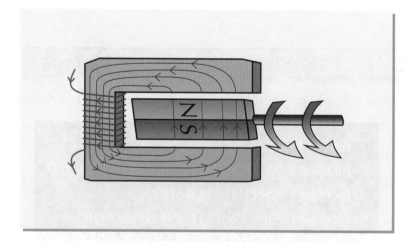

6) If the magnet keeps turning in the <u>same direction</u> — clockwise, say — then
the voltage keeps on reversing every half turn and you get an <u>AC current</u>.

So THAT's how they make electricity — I always wondered...

Generators are mostly powered by <u>burning things</u> to make <u>steam</u>, to turn a turbine, to rotate the magnet.
You can get portable generators too, to use in places without mains electricity — like at music festivals.

Transformers

Transformers are used to change the size of a voltage in a circuit — they're more useful than you think.

Transformers Change the Voltage — but Only AC Voltages

1) <u>Transformers</u> are used to change the <u>size</u> of the <u>voltage</u> —
   they use <u>electromagnetic induction</u> to 'step up' or 'step down' the <u>voltage</u>.

2) They have two coils of wire, the <u>primary</u> and the <u>secondary</u> coils, wound around an <u>iron core</u>.

3) The <u>alternating current</u> in the <u>primary</u> coil causes <u>changes</u> in the iron core's <u>magnetic field</u>,
   which <u>induces</u> a <u>changing voltage</u> in the <u>secondary</u> coil (see below).

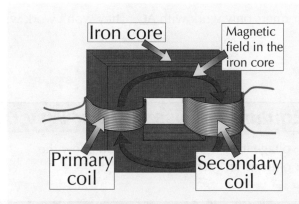

> **STEP-UP TRANSFORMERS** step the voltage **up** (increase it).
> They have **more** turns on the **secondary** coil than the primary coil.

> **STEP-DOWN TRANSFORMERS** step the voltage **down** (decrease it).
> They have **more** turns on the **primary** coil than the secondary.

Transformers Work by Electromagnetic Induction

1) The primary coil <u>produces a magnetic field</u> which stays <u>within the iron core</u>.

2) Because there's an <u>alternating current</u> (AC) in the <u>primary coil</u>,
   the magnetic field in the iron core constantly <u>changes direction</u>
   (100 times a second if it's at 50 Hz) — i.e. it's a <u>changing</u> magnetic field.

3) This changing magnetic field <u>induces</u> an <u>alternating voltage</u> in the secondary
   coil (with the same frequency as the alternating current in the primary) —
   <u>electromagnetic induction</u> of a voltage in fact.

Transformers

Transformers Need AC to Work

1) The <u>relative number of turns</u> on the two coils determines whether the voltage induced in the secondary coil is <u>greater</u> or <u>less</u> than the voltage in the primary coil (see equation below).

2) If you supplied <u>direct current</u> (DC) to the primary coil, you'd get <u>nothing</u> out of the secondary coil at all.

3) Sure, there'd still be a magnetic field in the iron core, but it wouldn't be <u>constantly changing</u>, so there'd be no <u>induction</u> in the secondary coil — because you need a <u>changing field</u> to induce a voltage.

4) So don't forget it — transformers only work with <u>AC</u>. They won't work with DC <u>at all</u>.

The Transformer Equation — use it Either Way Up

You can calculate the <u>output</u> voltage from a transformer if you know the <u>input</u> voltage and the number of turns on each coil.

Voltage across Primary Coil	=	Number of turns in Primary Coil
Voltage across Secondary Coil		Number of turns in Secondary Coil

Well, it's <u>just another formula</u>. You stick in the numbers <u>you've got</u> and work out the one <u>that's left</u>.

And you can write the formula <u>either way up</u> — you should always put the thing you're trying to find <u>on the top</u>.

$$\frac{V_P}{V_S} = \frac{N_P}{N_S}$$

or

$$\frac{V_S}{V_P} = \frac{N_S}{N_P}$$

Example

A transformer has 40 turns on the primary and 800 on the secondary. If the input voltage is 1000 V, find the output voltage.

<u>ANSWER</u>: The question asks you to find V_S, so put it on the top: $\quad \frac{V_S}{V_P} = \frac{N_S}{N_P}$

Substitute the values: $\quad \frac{V_S}{1000} = \frac{800}{40}, \quad V_S = 1000 \times \frac{800}{40} = \underline{20\ 000\ V}$

Transformers only work with AC

I'll say that again. Transformers only work with AC. Now that's out of the way — make sure you <u>practise</u>, <u>practise</u>, <u>practise</u> using that tricky equation... It's not your usual bog standard equation you know.

Magnetic Fields

Electric currents can create magnetic fields.

Magnetic Fields are Areas Where a Magnetic Force Acts

There's a proper definition of a magnetic field:

> A MAGNETIC FIELD is a region where MAGNETIC MATERIALS (like iron and steel) and also WIRES CARRYING CURRENTS experience A FORCE acting on them.

Magnetic fields can be represented by field diagrams (e.g. see coil of wire diagram below).

The arrows on the field lines always point FROM THE NORTH POLE of the magnet TO THE SOUTH POLE.

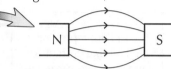

A Current-Carrying Wire Creates a Magnetic Field

1) When a current flows through a wire, a magnetic field is created around the wire.
2) The field is made up of concentric circles with the wire in the centre.

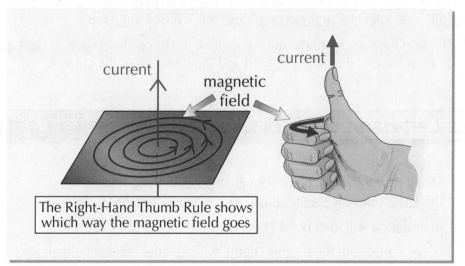

The Right-Hand Thumb Rule shows which way the magnetic field goes

A Rectangular Coil Reinforces the Magnetic Field

1) If you bend the current-carrying wire round into a coil, the magnetic field looks like this.

2) The circular magnetic fields around the sides of the loop reinforce each other at the centre.

3) If the coil has lots of turns, the magnetic fields from all the individual loops reinforce each other even more.

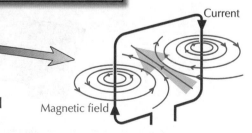

Just point your thumb in the direction of the current...

...and your fingers show the direction of the field. Remember, it's always your right thumb. Not your left, but your right thumb. You'll use your left hand on page 240 though, so it shouldn't feel left out...

Magnetic Fields

Passing an electric current through a wire produces a magnetic field around the wire (p.239).
If you put that wire into a magnetic field, you have <u>two magnetic fields combining</u>, which puts
a force on the wire (generally).

A *Current* in a *Magnetic Field* Experiences a *Force*

1) Because of its magnetic field, a <u>current-carrying wire</u> or <u>coil</u> can exert a <u>force</u> on
<u>another</u> current-carrying wire or coil, or on a <u>permanent magnet</u>.

2) When a current-carrying wire is put in a <u>different</u> magnetic field, the <u>two</u> magnetic
fields <u>affect one another</u>. The result is a <u>force</u> on the <u>wire</u>.

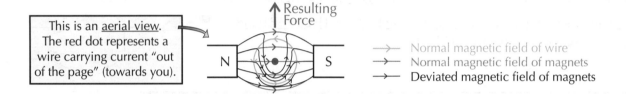

This is an <u>aerial view</u>.
The red dot represents a
wire carrying current "out
of the page" (towards you).

→ Normal magnetic field of wire
→ Normal magnetic field of magnets
→ Deviated magnetic field of magnets

3) To feel the <u>full force</u>, the <u>wire</u> has to be at <u>right-angles</u> (<u>90°</u>) to the <u>lines</u>
<u>of force</u> of the <u>magnetic field</u> it's placed in. (As it is in the diagram above.)

4) If the wire runs <u>parallel</u> to the lines of force of the magnetic field, it won't experience
<u>any force at all</u>. And at angles <u>in between</u> 0° and 90° it'll feel <u>some</u> force.

5) When the wire is at right-angles to the magnetic field, the <u>force</u> always acts at <u>right-angles</u>
to <u>both</u> the lines of force of the magnetic field <u>and</u> the direction of the current.

Fleming's Left-Hand Rule Tells You *Which Way* the *Force Acts*

Using your <u>left hand</u>, point your <u>First finger</u> in the direction of
the <u>Field</u> and your <u>seCond finger</u> in the direction of the <u>Current</u>.

Your <u>thuMb</u> will then point in the direction of the <u>force</u> (Motion).

(Give it a try with the diagram of the wire and the magnet field above.)

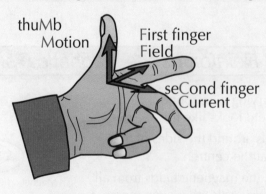

Thumb = Motion, First finger = Field, seCond finger = Current

I wonder if Fleming's left hand used to get worn out after a hard day at the office... But anyway, don't
just concentrate on the fun stuff at the bottom of this page — read through the stuff at the top too.

The Motor Effect

The <u>motor effect</u> — that's how <u>electric motors</u> work. Should be <u>easy</u> to remember that.

Magnetic Fields Make Current-Carrying Coils Turn

A uniform magnetic field has the same strength everywhere in the field.

If a <u>rectangular coil</u> of wire carrying a <u>current</u> is placed in a <u>uniform</u> magnetic field, the <u>force</u> will cause it to <u>turn</u>. This is called the <u>motor effect</u>. You can use Fleming's <u>left-hand rule</u> (<u>LHR</u>), from the previous page, to work out which <u>way</u> the coil will turn:

<u>EXAMPLE</u>: Is the coil turning clockwise or anticlockwise?

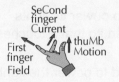

<u>ANSWER</u>:

1) Draw in current arrows (+ve to –ve) and magnetic field lines, which always run from North to South.

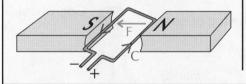

2) Use LHR on one side of the coil (I've used the right-hand side).

SeCond finger Current
First finger Field
thuMb Motion

3) Draw in direction of motion (force).

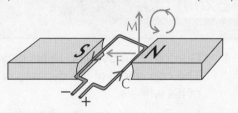

So — the coil is turning <u>anticlockwise</u>.

The Simple Electric Motor

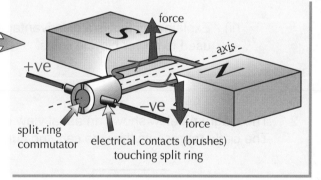

force
axis
+ve
–ve
force
split-ring commutator
electrical contacts (brushes) touching split ring

1) The diagram shows the <u>forces</u> acting on the two <u>side arms</u> of a <u>coil</u>.

2) These forces are just the <u>usual forces</u> which act on <u>any</u> current-carrying wire in a <u>magnetic field</u>.

3) Because the coil is on a <u>spindle</u> and the forces act <u>one up</u> and <u>one down</u>, it <u>rotates</u>.

4) The <u>split-ring commutator</u> is a clever way of <u>swapping</u> the contacts <u>every half turn</u>.

5) This reverses the direction of the <u>current</u> every half-turn to keep the coil rotating <u>continuously</u> in the <u>same direction</u>.

6) Otherwise, the direction of the <u>force</u> would <u>reverse</u> every half turn and the coil would <u>change direction</u> every half turn instead of fully <u>rotating</u>.

Anything That Uses Rotation can be Powered by an Electric Motor

Lots of devices use <u>rotation</u>. They all work by using an <u>electric motor</u> in a similar way.

1) Link the coil to an <u>axle</u>, and the axle <u>spins round</u>.

2) In the diagram there's a <u>fan</u> attached to the axle, but you can stick <u>almost anything</u> on a motor axle and make it spin round.

axle
fan
coil

3) For example:

- In a <u>DVD player</u>, the axle's attached to the bit <u>the DVD sits on</u> to make it spin.
- Electric <u>cars</u> and <u>trains</u> have their <u>wheels</u> attached to axles.
- Electric motors spin the <u>platters</u> (the bits where information is stored) of a computer <u>hard disc drive</u>.
- <u>Domestic appliances</u>, such as washing machines, fridges and vacuum cleaners use electric motors.

Warm-Up and Exam Questions

It's time for another page of questions to check your knowledge retention. If you can do the warm-up questions without breaking into a sweat, then see how you get on with the exam questions below.

Warm-Up Questions

1) What is meant by electromagnetic induction?
2) Do step-up transformers have more turns on their primary or secondary coil?
3) Write down the transformer equation.
4) In Fleming's left-hand rule, what's represented by the first finger? the second finger? the thumb?
5) Give three devices that are powered by an electric motor.

Exam Questions

1   Gordon fits a dynamo to his bicycle, to power its lights.
    The cog wheel of the dynamo is placed so that it touches the top of one
    of his wheels.

    (a) Describe how the dynamo is used to power the lights.

    (3 marks)

    (b) (i) Describe what happens to the dynamo's output when
              the wheel is not rotating.

    (1 mark)

         (ii) Explain why this is a disadvantage when a dynamo is
              used to power bicycle lights.

    (1 mark)

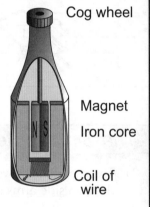

Cog wheel

Magnet

Iron core

Coil of wire

2   The diagram below shows an aerial view of a current-carrying wire in a magnetic field.
    The circle represents the wire carrying current out of the page, towards you.

    N   ◯   S

    (a) Describe the direction of the magnetic field of the magnet.

    (1 mark)

    (b) On the diagram, draw an arrow to show the direction of the force acting on the
         current-carrying wire.

    (1 mark)

    (c) Describe what would happen to the force acting on the current-carrying wire if the
         direction of the current was reversed.

    (1 mark)

    (d) Describe how the size of the force acting on the wire would change if:

         (i)   the wire was at 30° to the magnetic field.

    (1 mark)

         (ii) the wire ran parallel to the magnetic field.

    (1 mark)

Exam Questions

3   Julia is designing a toy car with a small electric motor to drive the wheels.
    The diagram shows a simplified version of Julia's motor.

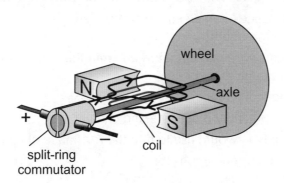

(a)   In which direction will the wheel turn?

(1 mark)

(b)   Explain the purpose of the split-ring commutator.

(1 mark)

(c)   Julia is testing her car and wants it to go faster.
      Suggest **one** change Julia can make to the motor to make the car go faster.

(1 mark)

(d)   Julia wants to make the car able to travel both forwards and backwards.
      Give **one** way she can reverse the direction of the wheels.

(1 mark)

4   A transformer is used to reduce the potential difference from a mains electricity supply.

(a)   Will the transformer have more turns on its primary or secondary coil?
      Explain your answer.

(1 mark)

(b)   Describe how a potential difference is induced in the secondary coil of the
      transformer.

(3 marks)

5   A student is trying to test a transformer using a battery as the power supply.

(a)   Explain why the voltmeter connected to the secondary coil reads 0 V.

(2 marks)

(b)   The student finds an AC power supply and reconnects the transformer.
      Her results are: V_P = 12 V, I_P = 2.5 A, V_S = 4 V (where V_P is the voltage across the
      primary coil, etc.).

   (i)   Calculate the power input to the transformer.

(2 marks)

   (ii) Calculate the current in the secondary coil, I_S. (Assume that power input to
         the transformer equals power output from the transformer.)

(3 marks)

   (iii) The primary coil has 15 turns.
         How many turns must be on the secondary coil?

(2 marks)

Revision Summary for Section 8

There's some pretty heavy physics in this section. But just take it one page at a time and it's not so bad.
You're even allowed to go back through the pages for a sneaky peak if you get stuck on these questions.

1)   What causes the build-up of static electricity? Which particles move when static builds up?

2)   Explain how static electricity can make synthetic clothes crackle when you take them off.

3)   True or false: the greater the resistance of an electrical component,
      the smaller the current that flows through it?

4)* 240 C of charge is carried though a wire in a circuit in one minute.
      How much current has flowed through the wire?

5)   What is another name for potential difference?

6)   What formula relates work done, potential difference and charge?

7)   Draw a diagram of the circuit that you would use to find the resistance of a lamp.

8)   Sketch typical potential difference-current graphs for:
      a) a resistor, b) a diode. Explain the shape of each graph.

9)   Explain how resistance of a component changes with its temperature in terms of ions and electrons.

10)* What potential difference is required to push 2 A of current through a 0.6 Ω resistor?

11)* Calculate the resistance of a wire if the potential difference across it is 12 V and the current through
      it is 2.5 A.

12) Give three applications of LEDs.

13) Describe how the resistance of an LDR varies with light intensity.

14) Explain, in terms of work done, why P.D. is shared out in a series circuit.

15) Two circuits each contain a 2 Ω and a 4 Ω resistor — in one circuit they're in series,
      in the other they're in parallel. Which circuit will have the higher total resistance? Why?

16)* An AC supply of electricity has a time period of 0.08s. What is its frequency?

17) Name the three wires in a three-core cable.

18) Sketch and label a properly wired three-pin plug.

19) How does an RCCB stop you from getting electrocuted?

20) What does the power of an appliance measure?

21)* Which uses more energy, a 45 W pair of hair straighteners used for 5 minutes,
      or a 105 W hair dryer used for 2 minutes?

22)* Calculate the energy transformed by a torch using a 6 V battery when 530 C of charge pass through.

23) What are the four factors that affect the size of the induced voltage produced by a generator?

24) Explain how a generator works — use a sketch if it helps.

25)* A transformer has 500 turns on the primary coil and 20 on the secondary coil.
      If the output voltage is 9 V, find the input voltage.

26) What part of an electric motor reverses the direction of the current? Why is this important?

* Answers on p.280.

Atomic Structure

Ernest Rutherford didn't just pick the nuclear model of the atom out of thin air. It all started with a Greek fella called Democritus in the 5th Century BC. He thought that <u>all matter</u>, whatever it was, was made up of <u>identical</u> lumps called "atomos". And that's about as far as the theory got until the 1800s...

Rutherford Scattering and the *Demise* of the *Plum Pudding*

1) In 1804 <u>John Dalton</u> agreed with Democritus that matter was made up of <u>tiny spheres</u> ("atoms") that couldn't be broken up, but he reckoned that <u>each element</u> was made up of a <u>different type</u> of "atom".

2) Nearly 100 years later, <u>J J Thomson</u> discovered that <u>electrons</u> could be <u>removed</u> from atoms. So Dalton's theory wasn't quite right (atoms could be broken up). Thomson suggested that atoms were <u>spheres of positive charge</u> with tiny negative electrons <u>stuck in them</u> like plums in a <u>plum pudding</u>.

3) That "plum pudding" theory didn't last very long though. In 1909 <u>Rutherford</u> and <u>Marsden</u> tried firing a beam of <u>alpha particles</u> (see p.249) at <u>thin gold foil</u>. They expected that the <u>positively charged</u> alpha particles would be <u>slightly deflected</u> by the widely spread positive charge in the plum pudding model.

4) However, most of the alpha particles just went <u>straight through</u>, but the odd one came <u>straight back</u> at them, which was frankly a bit of a <u>shocker</u> for Rutherford and his pal.

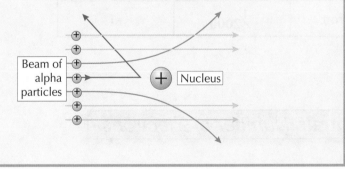

Beam of alpha particles

Nucleus

5) Being pretty clued-up guys, Rutherford and Marsden realised this meant that <u>most of the mass</u> of the atom was concentrated at the <u>centre</u> in a <u>tiny nucleus</u>. They also realised that the nucleus must have a <u>positive charge</u>, since it repelled the positive alpha particles.

6) It also showed that most of an atom is just <u>empty space</u>, which is also a bit of a <u>shocker</u> when you think about it.

Atomic Structure

Rutherford and Marsden used the results from their scattering experiment (see previous page) to produce a model for the atom.

Rutherford and Marsden Came Up with the Nuclear Model of the Atom

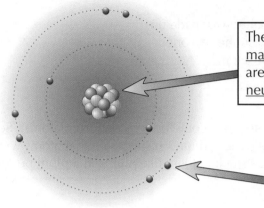

The nucleus is tiny but it makes up most of the mass of the atom. It contains protons (which are positively charged) and neutrons (which are neutral) — which gives it an overall positive charge.

The rest of the atom is mostly empty space.
The negative electrons whizz round the outside of the nucleus really fast. They give the atom its overall size — the radius of the atom's nucleus is about 10 000 times smaller than the radius of the atom.

We can use Relative Charges and Masses to Compare Particles

Here are the relative charges and relative masses of each particle:

Particle	Mass	Charge
Proton	1	+1
Neutron	1	0
Electron	$\frac{1}{2000}$	−1

Number of Protons Equals Number of Electrons

1) Atoms have no charge overall.

2) The charge on an electron is the same size as the charge on a proton — but opposite.

3) This means the number of protons always equals the number of electrons in a neutral atom.

4) If some electrons are added or removed, the atom becomes a charged particle called an ion.

The nuclear model is just one way of thinking about the atom

Rutherford and Marsden's nuclear model works really well for explaining a lot of physical properties of different elements — but it's certainly not the whole story. Other bits of science are explained using different models of the atom. The beauty of it though is that no one model is more right than the others.

Atoms and Radiation

You have just entered the subatomic realm — now stuff starts to get real interesting...

Isotopes are Different Forms of the Same Element

1) Isotopes are atoms with the same number of protons but a different number of neutrons.

2) Hence they have the same atomic number, but different mass numbers.

3) Atomic (proton) number is the number of protons in an atom.
   Mass (nucleon) number is the number of protons + the number of neutrons in an atom.

4) Carbon-12 and carbon-14 are good examples of isotopes:

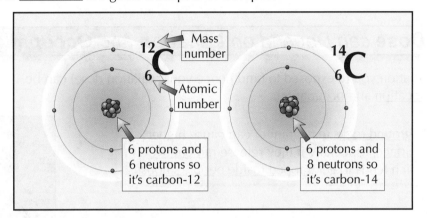

5) Most elements have different isotopes, but there are usually only one or two stable ones (like carbon-12).

6) The other isotopes tend to be radioactive (like carbon-14) which means they decay into other elements and give out radiation.

Radioactivity is a Totally Random Process

1) Radioactive substances give out radiation from the nuclei of their atoms — no matter what is done to them.

2) This process is entirely random. This means that if you have 1000 unstable nuclei, you can't say when any one of them is going to decay, and neither can you do anything at all to make a decay happen. It's completely unaffected by physical conditions like temperature or by any sort of chemical bonding etc.

3) Radioactive substances spit out one or more of the three types of radiation, alpha, beta or gamma (see page 249). In the process, the nucleus will often change into a new element.

Background Radiation Comes from Many Sources

Background radiation is radiation that is present at all times, all around us, wherever you go.
The background radiation we receive comes from:

1) Radioactivity of naturally occurring unstable isotopes which are all around us — in the air, in food, in building materials and in the rocks under our feet.

2) Radiation from space, which is known as cosmic rays. These come mostly from the Sun.

3) Radiation due to man-made sources, e.g. fallout from nuclear weapons tests, nuclear accidents (such as Chernobyl) or dumped nuclear waste.

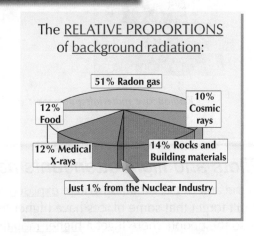

The RELATIVE PROPORTIONS of background radiation:

Radiation Dose

This page is all about what <u>affects</u> the amount of <u>radiation</u> we're <u>exposed</u> to, and the damage it does.

The **Damage** Caused by **Radiation Depends** on the **Dose**

How likely you are to <u>suffer damage</u> if you're exposed to nuclear radiation depends on the <u>radiation dose</u>.

1) Radiation dose is measured in sieverts (Sv) and depends on the <u>type</u> and <u>amount of radiation</u> you've been exposed to.

2) The <u>higher</u> the radiation dose, the <u>more at risk</u> you are of <u>developing cancer</u>.

Radiation Dose can Depend on **Location** and **Occupation**

The amount of radiation you're exposed to (and hence your radiation dose) can be affected by your <u>location</u> and <u>occupation</u>:

1) Certain <u>underground rocks</u> (e.g. granite) can cause higher levels at the <u>surface</u>, especially if they release <u>radioactive radon gas</u>, which tends to get <u>trapped inside people's houses</u>.

Coloured bits indicate more radiation from rocks

2) <u>Nuclear industry</u> workers and <u>uranium miners</u> have a higher risk of radiation exposure. They wear <u>protective clothing</u> and <u>face masks</u> to stop them from <u>touching</u> or <u>inhaling</u> the radioactive material, and <u>monitor</u> their radiation doses with <u>special radiation badges</u> and <u>regular check-ups</u>.

3) <u>Radiographers</u> work in hospitals using ionising radiation and so have a higher risk of radiation exposure. They wear <u>lead aprons</u> and stand behind <u>lead screens</u> to protect them from <u>prolonged exposure</u> to radiation.

4) At <u>high altitudes</u> (e.g. in <u>jet planes</u>) the background radiation <u>increases</u> because of more exposure to <u>cosmic rays</u>. That means <u>commercial pilots</u> have an increased risk of getting some types of cancer.

5) <u>Underground</u> (e.g. in <u>mines</u>, etc.) it increases because of the <u>rocks</u> all around, posing a risk to <u>miners</u>.

Radioactive materials put people at risk through either <u>irradiation</u> or <u>contamination</u> (see p.254).

Pilots and flight attendants have a greater exposure to cosmic rays

So the amount of radiation you're <u>exposed</u> to depends on your <u>job</u> and your <u>location</u>. Don't forget that some places have higher levels of background radiation than others — so the people there'll get a <u>higher radiation dose</u>.

Ionising Radiation

Alpha (α) Beta (β) Gamma (γ) — there's a short alphabet of radiation for you to learn. And it's all _ionising_.

Alpha Particles are *Helium Nuclei*

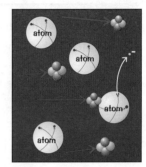

Alpha particles have a mass of 4 and a charge of +2 — just like helium nuclei.

1) An <u>alpha</u> particle is <u>two neutrons</u> and <u>two protons</u>.

2) They are relatively <u>big</u> and <u>heavy</u> and <u>slow moving</u>.

3) They therefore <u>don't</u> penetrate very far into materials and are <u>stopped quickly</u>, even when travelling through <u>air</u>.

4) Because of their size they are <u>strongly ionising</u>, which just means they <u>bash into</u> a lot of atoms and <u>knock electrons off them</u> before they slow down, which creates lots of ions — hence the term "<u>ionising</u>".

Beta Particles are *Electrons*

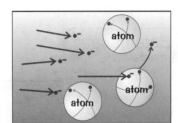

1) Beta particles are <u>in between</u> alpha and gamma in terms of their <u>properties</u>.

2) They move <u>quite</u> fast and they are <u>quite</u> small (they're electrons).

3) They <u>penetrate moderately</u> into materials before colliding, have a <u>long range</u> in air, and are <u>moderately ionising</u> too.

4) For every β-particle emitted, a <u>neutron</u> turns to a <u>proton</u> in the nucleus.

5) A <u>β-particle</u> is simply an <u>electron</u>, with virtually no mass and a charge of −1.

Gamma Rays are Very Short Wavelength *EM Waves*

1) If nuclei need to get rid of some <u>extra energy</u> they can emit <u>gamma rays</u>.

2) Gamma rays are just <u>energy</u> (they have <u>no mass</u> and <u>no charge</u>) so they <u>don't</u> change the element of the nucleus that emits it.

3) They can <u>penetrate a long way</u> into materials without being stopped and pass <u>straight through air</u>.

4) This means they are <u>weakly</u> ionising because they tend to <u>pass through</u> rather than collide with atoms. Eventually they <u>hit something</u> and do <u>damage</u>.

Alpha and beta emissions are particles, gamma emissions are rays

So, when a nucleus decays by <u>alpha</u> emission, its atomic number <u>goes down</u> by <u>two</u> and its mass number goes down by <u>four</u>. <u>Beta</u> emission <u>increases</u> the atomic number by <u>one</u> (the mass number doesn't change). Gamma emission is just energy so it doesn't change the nucleus that emits it.

Ionising Radiation

When nuclei decay by <u>alpha</u> or <u>beta</u> emission, they change from one element into a different one.

Nuclear Equations *Need to* *Balance*

1) You can write alpha and beta decays as <u>nuclear equations</u>.

2) Watch out for the <u>mass and atomic numbers</u> — they have to <u>balance up</u> on both sides.

Alpha decay:

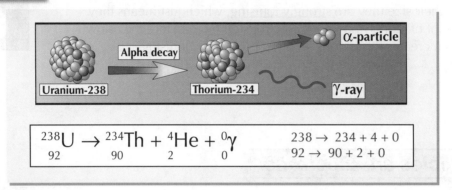

$$^{238}_{92}U \rightarrow \,^{234}_{90}Th + \,^{4}_{2}He + \,^{0}_{0}\gamma$$

$$238 \rightarrow 234 + 4 + 0$$
$$92 \rightarrow 90 + 2 + 0$$

Beta decay:

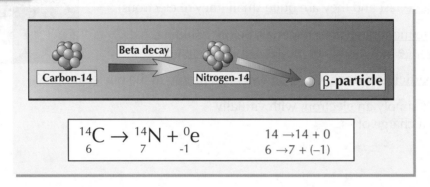

$$^{14}_{6}C \rightarrow \,^{14}_{7}N + \,^{0}_{-1}e$$

$$14 \rightarrow 14 + 0$$
$$6 \rightarrow 7 + (-1)$$

Alpha and *Beta* are *Deflected* by *Electric* and *Magnetic Fields*

1) Alpha particles have a <u>positive charge</u>, beta particles have a <u>negative charge</u>.

2) When travelling through a <u>magnetic</u> or <u>electric field</u>, both alpha and beta particles will be <u>deflected</u>.

3) They're deflected in <u>opposite directions</u> because of their <u>opposite charge</u>.

4) Alpha particles have a <u>larger charge</u> than beta particles, and feel a <u>greater force</u> in magnetic and electric fields. But they're <u>deflected less</u> because they have a <u>much greater mass</u>.

5) <u>Gamma radiation</u> is an electromagnetic (EM) wave and has <u>no charge</u>, so it <u>doesn't get</u> <u>deflected</u> by electric or magnetic fields.

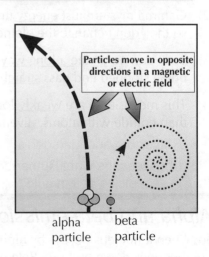

Particles move in opposite directions in a magnetic or electric field

alpha particle   beta particle

Warm-Up and Exam Questions

It's time again to test what you've learnt from the last few pages. Have a go at these...

Warm-Up Questions

1) In the 'plum pudding' model of the atom, what are the 'plums'?
2) Give one man-made source of background radiation.
3) Give two factors that affect a person's average yearly radiation dose.
4) Which type of background radiation are pilots more exposed to than the average person?
5) Which are the most ionising — alpha particles or gamma rays?

Exam Questions

1 (a) Give the relative charge of the following particles:

    (i) electron

    (ii) proton

    (iii) neutron

(3 marks)

(b) Name the two types of particle that the nucleus of an atom contains.

(2 marks)

(c) Describe how the atomic number of a nucleus changes when a beta particle is emitted.

(1 mark)

(d) Describe how the mass number of a nucleus changes after alpha emission.

(1 mark)

(e) The table below contains information about three atoms.

	Mass number	Atomic number
Atom A	32	17
Atom B	33	17
Atom C	32	16

    (i) What is meant by the mass number of an atom?

(1 mark)

    (ii) Which of the two atoms are isotopes of the same element? Explain your answer.

(2 marks)

(f) Alpha and beta particles are deflected in electric and magnetic fields.

    (i) Explain why alpha and beta particles are deflected in opposite directions.

(1 mark)

    (ii) Explain why alpha particles are deflected less than beta particles.

(1 mark)

2 Rutherford and Marsden's scattering experiment led to their nuclear model of the atom.

(a) Describe what Rutherford and Marsden saw when they fired a beam of alpha particles at thin gold foil.

(2 marks)

(b) Describe the main features of their nuclear model of the atom.

(4 marks)

Half-Life

The <u>unit</u> for measuring <u>radioactivity</u> is the <u>becquerel</u> (Bq). 1 Bq means <u>one nucleus decaying per second</u>.

The **Radioactivity** of a Sample Always **Decreases** over Time

1) Each time a <u>decay</u> happens and an alpha, beta or gamma is given out, it means one more <u>radioactive nucleus</u> has <u>disappeared</u>.

2) Obviously, as the <u>unstable nuclei</u> all steadily disappear, the <u>activity</u> (the number of nuclei that decay per second) will <u>decrease</u>. So the <u>older</u> a sample becomes, the <u>less radiation</u> it will emit.

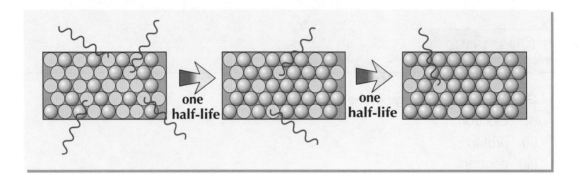

3) <u>How quickly</u> the activity <u>drops off</u> varies a lot. For <u>some</u> substances it takes <u>just a few microseconds</u> before nearly all the unstable nuclei have <u>decayed</u>, whilst for others it can take <u>millions of years</u>.

4) The problem with trying to <u>measure</u> this is that <u>the activity never reaches zero</u>, which is why we have to use the idea of <u>half-life</u> to measure how quickly the activity <u>drops off</u>.

5) Here's the <u>definition</u> of <u>half-life</u>:

> **HALF-LIFE is the <u>AVERAGE TIME</u> it takes for the <u>NUMBER OF NUCLEI</u> in a <u>RADIOACTIVE ISOTOPE SAMPLE</u> to <u>HALVE</u>.**

6) In other words, it is the <u>time it takes</u> for the <u>count rate</u> (the number of radioactive emissions detected per unit of time) from a sample containing the isotope to <u>fall to half its initial level</u>.

7) A <u>short half-life</u> means the <u>activity falls quickly</u>, because <u>lots</u> of the nuclei decay <u>quickly</u>.

8) A <u>long half-life</u> means the activity <u>falls more slowly</u> because <u>most</u> of the nuclei don't decay <u>for a long time</u> — they just sit there, <u>basically unstable</u>, but kind of <u>biding their time</u>.

Half-life measures how quickly the activity of a source drops off

For <u>medical applications</u>, you need to use isotopes that have a <u>suitable half-life</u>.
A radioactive tracer needs to have a short half-life to minimise the risk of damage to the patient.
A radioactive source for sterilising equipment needs to have a long half-life, so you don't have to replace it too often (see page 257 for more). There's more on half-life coming up...

Half-Life

This page is about <u>how to tackle</u> the two main types of half-life questions.

Do *Half-Life* Questions *Step by Step*

Half-life can be confusing, but the calculations are <u>straightforward</u> so long as you do them <u>STEP BY STEP</u>:

A Very Simple Example:

The activity of a radioactive sample is 640 Bq.
Two hours later it has fallen to 40 Bq. Find its half-life.

<u>ANSWER:</u> Go through it in <u>short simple steps</u> like this:

INITIAL		after **ONE**		after **TWO**		after **THREE**		after **FOUR**
count:	(÷2) →	half-life:	(÷2) →	half-lives:	(÷2) →	half-lives:	(÷2) →	half-lives:
640		**320**		**160**		**80**		**40**

This careful <u>step-by-step method</u> shows that it takes
<u>four half-lives</u> for the activity to fall from 640 to 40.

So <u>two hours</u> represents <u>four half-lives</u> — so the half-life is 2 hours ÷ 4 = **30 MINUTES**.

And here's how you calculate the half-life of a sample from a <u>graph</u>. Relax, this is (almost) <u>fun</u>.

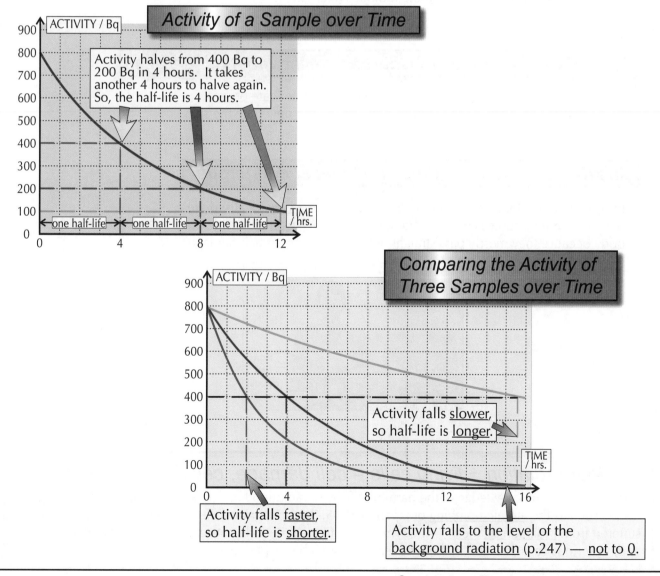

Activity of a Sample over Time

Activity halves from 400 Bq to 200 Bq in 4 hours. It takes another 4 hours to halve again. So, the half-life is 4 hours.

one half-life | one half-life | one half-life

Comparing the Activity of Three Samples over Time

Activity falls <u>slower</u>, so half-life is <u>longer</u>.

Activity falls <u>faster</u>, so half-life is <u>shorter</u>.

Activity falls to the level of the <u>background radiation</u> (p.247) — <u>not</u> to <u>0</u>.

Radioactivity Safety

When <u>Marie Curie</u> discovered the radioactive properties of <u>radium</u> in 1898, nobody knew about its dangers. Radium was used to make glow-in-the-dark watches and many <u>watch dial painters</u> developed cancer as a result. We now know lots more about the dangers of radiation...

Ionising Radiation Can Damage Living Cells

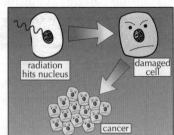

radiation hits nucleus — damaged cell — cancer

1) Alpha, beta and gamma radiation are all <u>ionising radiation</u> — they can <u>break up</u> molecules into smaller bits called <u>ions</u>.

2) Ions can be <u>very chemically reactive</u>, so they go off and react with things and generally make <u>nuisances</u> of themselves.

3) In humans, ionisation can cause <u>serious damage</u> to the cells in the body.

4) A high dose of radiation tends to <u>kill cells</u> outright, causing <u>radiation sickness</u>.

5) Lower doses tend to <u>damage cells</u> without killing them, which can cause <u>cancer</u>.

6) Radioactive materials put people at risk through either:

   - <u>IRRADIATION</u> — being exposed to radiation <u>without</u> coming into contact with the source. The damage to your body <u>stops</u> as soon as you leave the radioactive area.
   - <u>CONTAMINATION</u> — <u>picking up</u> some radioactive material, e.g. by <u>breathing it in</u>, <u>drinking</u> contaminated water or getting it on your skin. You'll <u>still</u> be exposed to the radiation once you've <u>left</u> the radioactive area.

Outside the body, β and γ–sources are the most dangerous

This is because <u>beta and gamma</u> can get <u>inside</u> to the delicate <u>organs</u>, whereas alpha is much less dangerous because it <u>can't penetrate the skin</u>.

Inside the body, an α-source is the most dangerous

<u>Inside the body</u> alpha sources do all their damage in a <u>very localised area</u>. Beta and gamma sources on the other hand are <u>less dangerous</u> inside the body because they mostly <u>pass straight out</u> without doing much damage.

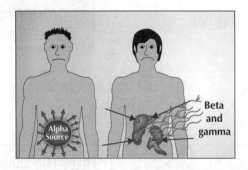

Alpha Source

Beta and gamma

Nuclear radiation + living cells = cell damage, cell death or cancer

Sadly, much of our knowledge of the harmful effects of radiation has come as a result of devastating events such as the <u>atomic bombing</u> of Japan in 1945. In the months following the bombs, <u>thousands</u> suffered from <u>radiation sickness</u> — the symptoms of which include nausea, fatigue, skin burns, hair loss and, in serious cases, death. In the long term, the area has experienced <u>increased rates</u> of cancer.

Radioactivity Safety

The last page was all about the damage that radiation can do to your body. Now here are some ways you can <u>protect</u> yourself <u>against exposure</u> to radiation.

You Need to Keep *Safe* whilst *Handling Radioactive Sources*

1) When conducting experiments, use radioactive sources for as <u>short a time</u> as possible so your <u>exposure</u> is kept to a <u>minimum</u>.

2) <u>Never</u> allow <u>skin contact</u> with a source. Always handle with <u>tongs</u>.

3) Hold the source at <u>arm's length</u> to keep it <u>as far</u> from the body <u>as possible</u>. This will decrease the amount of radiation that hits you, especially for alpha particles as they <u>don't travel far in air</u>.

4) Keep the source <u>pointing away</u> from the body and <u>avoid looking directly at it</u>.

Here's What *Blocks* the *Three Types* of *Radiation...*

1) <u>Alpha particles</u> are blocked by <u>paper</u>.
2) <u>Beta particles</u> are blocked by thin <u>aluminium</u>.
3) <u>Gamma rays</u> are blocked by <u>thick lead</u>.
4) <u>Similar</u> things will also block them, e.g. <u>skin</u> will stop <u>alpha</u>, a thin sheet of <u>any metal</u> will stop <u>beta</u>, and <u>very thick concrete</u> will stop <u>gamma</u>.

Lead can Help *Protect* us from *Exposure* to *Radiation*

1) <u>Lead</u> absorbs all three types of radiation (though a lot of it is needed to stop gamma radiation completely). <u>Always</u> store radioactive sources in a <u>lead box</u> and put them away <u>as soon</u> as the experiment is <u>over</u>.

2) Medical professionals who work with radiation <u>every day</u> (such as radiographers) wear <u>lead aprons</u> and stand behind <u>lead screens</u> for extra protection because of its radiation absorbing properties.

3) When someone needs an X-ray or radiotherapy, only the area of the body that <u>needs to be treated</u> is exposed to radiation. The rest of the body is <u>protected with lead</u> or other <u>radiation absorbing</u> materials.

Radiation's dangerous stuff — safety precautions are crucial

Radiation can be harmful to us (see previous page) but following the safety precautions above can <u>minimise</u> our exposure to radiation. This page gives you loads of information about how to <u>protect</u> yourself in the laboratory, and how <u>medical workers</u> can protect themselves from radiation too.

Uses of Radiation

Radiation gets a lot of bad press, but the fact is it's essential for things like <u>modern medicine</u>.

Smoke Detectors — Use α-Radiation

1) A <u>weak alpha</u> radioactive source is placed in the detector, close to <u>two electrodes</u>.
2) The source causes <u>ionisation</u> of the air particles which allows a <u>current</u> to flow.
3) If there is a fire, then <u>smoke particles</u> are hit by the alpha particles instead.
4) This causes <u>less ionisation</u> of the air particles — so the <u>current</u> is <u>reduced</u> causing the <u>alarm</u> to <u>sound</u>.

Radiotherapy — the Treatment of Cancer Using γ-Rays

1) Since high doses of gamma rays will <u>kill all living cells</u>, they can be used to <u>treat cancers</u>.
2) The gamma rays have to be <u>directed carefully</u> and at just the right <u>dosage</u> so as to kill the <u>cancer cells</u> without damaging too many <u>normal cells</u>.
3) However, a <u>fair bit of damage</u> is <u>inevitably</u> done to <u>normal cells</u>, which makes the patient feel <u>very ill</u>. But if the cancer is <u>successfully killed off</u> in the end, then it's worth it.

TO TREAT CANCER:
1) The gamma rays are <u>focused</u> on the tumour using a <u>wide beam</u>.
2) This beam is <u>rotated</u> round the patient with the tumour at the centre.
3) This <u>minimises</u> the exposure of <u>normal cells</u> to radiation, and so <u>reduces</u> the chances of damaging the rest of the body.

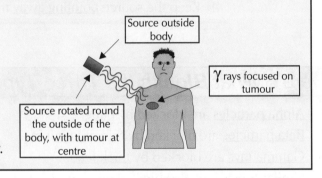

Source outside body

γ rays focused on tumour

Source rotated round the outside of the body, with tumour at centre

Tracers in Medicine — Always Short Half-life β or γ -Emitters

1) Certain <u>radioactive isotopes</u> that emit <u>gamma</u> (and sometimes <u>beta</u>) radiation can be used as <u>tracers</u> in the body.
2) They should have a <u>short half-life</u> — around <u>a few hours</u>, so that the radioactivity inside the patient <u>quickly disappears</u>.
3) They can be <u>injected</u> inside the body, <u>drunk</u> or <u>eaten</u> or <u>ingested</u>.
4) They are allowed to <u>spread</u> through the body and their progress can be followed on the outside using a <u>radiation detector</u>.

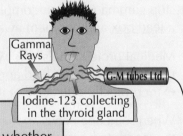

Gamma Rays

G-M tubes Ltd.

Iodine-123 collecting in the thyroid gland

Example <u>Iodine-123</u> is absorbed by the <u>thyroid gland</u>. It gives out <u>radiation</u> which can be <u>detected</u> to indicate whether or not the thyroid gland is <u>taking in the iodine</u> as it should.

5) <u>All isotopes</u> which are taken <u>into the body</u> must be <u>gamma or beta</u> (never alpha). This is because gamma and beta radiation can <u>penetrate tissue</u> and so are able to <u>pass out of the body</u> and be <u>detected</u>.
4) Alpha radiation <u>can't penetrate tissue</u>, so you <u>couldn't detect</u> the radiation on the outside of the body. Also alpha is more dangerous <u>inside the body</u> (see page 254).

Uses of Radiation

Tracers in **Industry** — For Finding **Leaks**

1) This is much the same technique as the medical tracers (see previous page).

2) Radioactive isotopes can be used to <u>track</u> the <u>movement</u> of <u>waste</u> materials, find the <u>route</u> of underground pipe systems or <u>detect leaks</u> or <u>blockages</u> in <u>pipes</u>.

3) To check a pipe, you just <u>squirt</u> the radioactive isotope in, then go along the <u>outside</u> with a <u>detector</u>. If the radioactivity <u>reduces</u> or <u>stops</u> after a certain point, there must be a <u>leak</u> or <u>blockage</u> there. This is really useful for <u>concealed</u> or <u>underground</u> pipes — no need to <u>dig up the road</u> to find the leak.

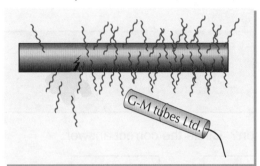

4) The isotope used <u>must</u> be a <u>gamma emitter</u>, so that the radiation can be <u>detected</u> even through <u>metal or earth</u> which may be <u>surrounding</u> the pipe. Alpha and beta radiation wouldn't be much use because they are <u>easily blocked</u> by any surrounding material.

5) It should also have a <u>short half-life</u> so as not to cause a <u>hazard</u> if it collects somewhere.

Sterilisation of **Food** and **Surgical Instruments** Using γ-Rays

1) <u>Food</u> can be exposed to a <u>high dose</u> of gamma rays which will <u>kill</u> all <u>microbes</u>, keeping the food <u>fresh for longer</u>.

2) <u>Medical instruments</u> can be <u>sterilised</u> in just the same way, rather than by <u>boiling them</u>.

3) The great <u>advantage</u> of <u>irradiation</u> over boiling is that it doesn't involve <u>high temperatures</u>, so things like <u>fresh apples</u> or <u>plastic instruments</u> can be totally <u>sterilised</u> without <u>damaging</u> them.

4) The food is <u>not</u> radioactive afterwards, so it's <u>perfectly safe</u> to eat.

5) The isotope used for this needs to be a <u>very strong</u> emitter of <u>gamma rays</u> with a <u>reasonably long half-life</u> (at least several months) so that it doesn't need <u>replacing</u> too often.

Beta Radiation is Used in **Thickness Gauges**

1) <u>Beta radiation</u> is used in <u>thickness control</u>.

2) You direct radiation through the stuff being made (e.g. paper), and put a detector on the other side, connected to a control unit.

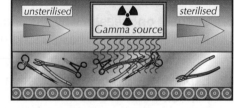

3) When the amount of <u>detected</u> radiation changes, it means the paper is coming out too thick or too thin, so the control unit adjusts the rollers to give the correct thickness.

4) The radioactive source used needs to have a fairly long half-life so it doesn't decay away <u>too quickly</u>.

5) It also needs to be a <u>beta</u> source, because then the paper will <u>partly block</u> the radiation (see p.255). If it <u>all</u> goes through (or <u>none</u> of it does), then the reading <u>won't change</u> at all as the thickness changes.

Make sure you choose the right radiation for the job

The different types of radiation (alpha, beta and gamma) all have different properties. This makes them useful for different things — so it's important to choose the type of radiation you use wisely.

Warm-Up and Exam Questions

There's no point in skimming through the section and glancing over the questions. Do the warm-up questions and go back over any bits you don't know. Then try the exam questions — without cheating.

Warm-Up Questions

1) Why is nuclear radiation dangerous?
2) Explain what irradiation is.
3) Why do medical professionals who work with radiation wear lead aprons?
4) How do smoke detectors work?

Exam Questions

1 Which of the following is not a use of gamma radiation? Circle the correct answer.

| treatment of cancer | smoke detectors | sterilising machines | medical tracers |

(1 mark)

2 Nuclear radiation has many uses within medicine.

(a) Suggest **two** reasons why alpha sources aren't used as medical tracers.

(2 marks)

(b) Suggest **one** reason why it is important that radioactive sources used in hospital sterilising machines have a long half-life.

(1 mark)

(c) Explain why the dose of radiation given in radiotherapy is directed only at the tumour.

(1 mark)

3 Nuclear radiation can have harmful effects on the human body.

(a) Briefly explain how a low dose of nuclear radiation can cause cancer.

(1 mark)

(b) Describe what can happen to the body if it receives a very high dose of nuclear radiation.

(1 mark)

4 A sample of a highly ionising radioactive gas has a half-life of two minutes.

(a) Describe what is meant by the term 'half-life'.

(1 mark)

(b) The sample contains a number of unstable atoms.
Calculate the fraction of these atoms that will be present after four minutes.

(1 mark)

(c) A worker holds a sample of the gas in a container using tongs.
Suggest **two** other ways she could protect herself against exposure to radiation from the gas.

(2 marks)

Exam Questions

5   A thick block of metal has been exposed to three different types of radiation. The numbered lines on the diagram show the distance each type of radiation penetrated into the block before being stopped.

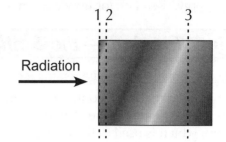

Radiation

   (a)   Which number corresponds to each of alpha, beta and gamma radiation?

(1 mark)

   (b)   Which type of radiation is the most dangerous inside the body? Explain your answer.

(2 marks)

6   Iodine-123 is used as a medical tracer.

   (a)   What organ is iodine-123 used to examine?

(1 mark)

   (b)   What type of radiation does iodine-123 emit?

(1 mark)

7   The diagram below shows the sterilisation of medical equipment by irradiation.

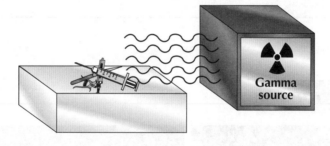

   (a)   Would a radioactive source with a very short half-life be used for this purpose? Explain your answer.

(1 mark)

   (b)   Explain the main advantage of irradiating equipment rather than boiling it.

(2 marks)

8   The diagram shows paper being made in a mill. If the thickness of the paper changes, so does the amount of beta radiation reaching the detector. The machinery then adjusts the rollers accordingly until the paper is the correct thickness.

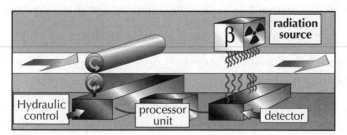

   (a)   Why isn't an alpha source used with this machinery?

(1 mark)

   (b)   Why isn't gamma radiation used for this purpose?

(1 mark)

Nuclear Fission

Unstable isotopes aren't just good for medicine — with the right set-up you can generate some <u>serious energy</u>. Read on for how we can use that energy in power stations...

Nuclear Fission — the *Splitting Up* of Big Atomic Nuclei

1) <u>Nuclear power stations</u> generate electricity using <u>nuclear reactors</u>.

2) In a nuclear reactor, a controlled <u>chain reaction</u> takes place in which atomic nuclei <u>split up</u> and <u>release energy</u> in the form of <u>heat</u>. This heat is then simply used to <u>heat water</u> to make steam, which is used to drive a <u>steam turbine</u> connected to an <u>electricity generator</u>.

3) The "<u>fuel</u>" that's split is usually <u>uranium-235</u>, though sometimes it's <u>plutonium-239</u> (or both).

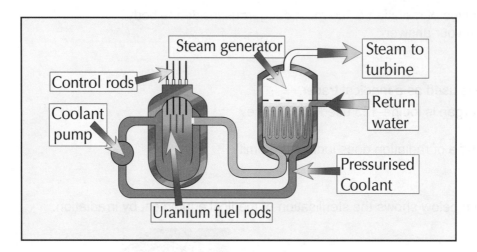

The *Chain Reactions*:

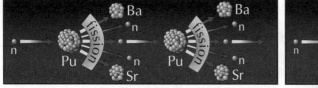 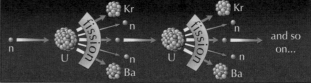

1) For nuclear fission to happen, a <u>slow moving neutron</u> must be <u>absorbed</u> into a uranium or plutonium nucleus. This addition of a neutron makes the nucleus unstable, causing it to <u>split</u>.

2) Each time a <u>uranium</u> or <u>plutonium</u> nucleus <u>splits up</u>, it spits out <u>two or three neutrons</u>, one of which might hit <u>another</u> nucleus, causing it to <u>split</u> also, and thus keeping the <u>chain reaction</u> going.

3) The chain reaction in the reactor has to be <u>controlled</u>, or the reactor would overheat.

4) <u>Control rods</u> <u>absorb</u> some of the <u>neutrons</u> and <u>slow down</u> the reaction.

5) When a large atom splits in two it will form <u>two new smaller nuclei</u>. These new nuclei are usually <u>radioactive</u> because they have the "<u>wrong</u>" number of neutrons in them.

6) A nucleus <u>splitting</u> (called a <u>fission</u>) gives out <u>a lot of energy</u> — lots more energy than you get from any <u>chemical</u> reaction. <u>Nuclear processes</u> release <u>much more energy</u> than chemical processes do. That's why <u>nuclear bombs</u> are <u>so much</u> more <u>powerful</u> than ordinary bombs (which rely on <u>chemical</u> reactions).

Nuclear Fission

At first sight, nuclear fission seems to be the perfect source of energy, but there are big problems with using nuclear power too — getting rid of nuclear waste can be a bit of an issue.

Nuclear Power Stations May Cause Quite a Few Problems

The main problem with nuclear power is that it produces radioactive waste...

1) Most waste from power stations (or medical use, see p.256) is 'low level' (slightly radioactive). E.g. things like paper and gloves, etc. This waste can be disposed of by burying it in secure landfill sites.

2) Intermediate level waste includes things like the metal cases of used fuel rods and some waste from hospitals. It's usually quite radioactive — and some of it will stay that way for tens of thousands of years. It's often sealed into concrete blocks then put in steel canisters for storage.

3) High level waste from nuclear power stations is so radioactive that it generates a lot of heat. This waste is sealed in glass and steel, then cooled for about 50 years before it's moved to more permanent storage.

4) The canisters of intermediate and high level wastes could then be buried deep underground. However, it's difficult to find suitable places. The site has to be geologically stable (e.g. not suffer earthquakes), since big movements in the rock could break the canisters and radioactive material could leak out. Even when geologists do find suitable sites, people who live nearby often object.

5) So, at the moment, most intermediate and high level waste is kept 'on-site' at nuclear power stations.

6) Nuclear fuel is cheap but the overall cost of nuclear power is high due to the cost of the power plant, waste processing and final decommissioning. Dismantling a nuclear plant safely takes decades.

7) Nuclear power also carries the risk of radiation leaks from the plant or a major catastrophe like Chernobyl.

Nuclear power — a replacement for fossil fuels?

There's plenty of plutonium and uranium around so some people think that nuclear power may one day replace fossil fuels as our main source of energy. But the problem of getting rid of nuclear waste is a big one. I bet you wouldn't want nuclear waste buried under your backyard, eh...

Nuclear Fusion

Scientists have been looking into producing energy the same way stars do — through <u>fusion</u>.

Nuclear Fusion — The *Joining* of Small Atomic Nuclei

1) <u>Nuclear fusion</u> is the <u>opposite</u> of nuclear <u>fission</u>.

2) In nuclear fusion, two <u>light nuclei combine</u> to create a larger nucleus.

3) One example is <u>two atoms</u> of different <u>hydrogen</u> isotopes combining to form <u>helium</u>:

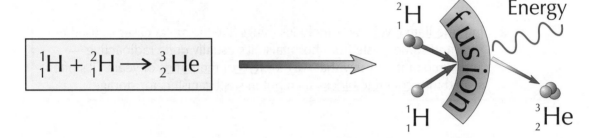

$$^{1}_{1}H + {}^{2}_{1}H \rightarrow {}^{3}_{2}He$$

4) Fusion releases <u>a lot</u> of energy (<u>more</u> than fission for a given mass) — all the energy released in <u>stars</u> comes from fusion at extremely <u>high temperatures</u> and <u>pressures</u>. So people are trying to develop <u>fusion reactors</u> to make <u>electricity</u>.

5) Fusion <u>doesn't</u> leave behind much radioactive <u>waste</u> and there's <u>plenty</u> of hydrogen about to use as <u>fuel</u>.

6) The <u>big problem</u> is that fusion only happens at <u>really high pressures</u> and <u>temperatures</u> (about <u>10 000 000 °C</u>). It doesn't happen at low temperatures and pressures due to the electrostatic repulsion of protons (like charges repel each other, see p.210).

7) <u>No material</u> can physically withstand that kind of temperature and pressure — so fusion reactors are <u>really hard</u> to <u>build</u>.

8) It's also hard to <u>safely control</u> the high temperatures and pressures.

9) There are a few <u>experimental</u> reactors around at the moment, the biggest one being <u>JET</u> (Joint European Torus), but <u>none</u> of them are <u>generating electricity yet</u>. It takes <u>more</u> power to get up to temperature than the reactor can produce.

10) <u>Research</u> into fusion power production is carried out by <u>international</u> groups to <u>share</u> the <u>costs</u>, <u>expertise</u>, experience and the <u>benefits</u> (when they eventually get it to work reliably).

Ten million degrees — that's hot...

It'd be great if we could get nuclear fusion to work — there's loads of fuel available and it doesn't create much radioactive waste compared with fission. It's a shame that at the moment we need to use more energy to create the conditions for fusion than we can get out of it.

Nuclear Fusion

Nuclear fusion has lots of potential — the problem is it requires really high temperatures.
But if we could get nuclear fusion to happen at room temperature — well, the possibilities are endless.

Fusion Bombs

1) Fusion reactions also happen in <u>fusion bombs</u>.

2) You might have heard of them as <u>hydrogen</u>, or <u>H bombs</u>.

3) In fusion bombs, a <u>fission reaction</u> is used first to create the really <u>high temperatures</u> needed for fusion.

Cold Fusion — Hoax or Energy of the Future?

1) A new scientific theory has to go through a <u>validation</u> process before it's accepted.

2) An example of a theory which <u>hasn't</u> been accepted yet is '<u>cold fusion</u>'.

3) Cold fusion is <u>nuclear fusion</u> which occurs at around <u>room temperature</u>, rather than at millions of degrees Celsius.

4) In 1989, two scientists reported that they had succeeded in releasing energy from cold fusion, using a simple experiment. This caused a lot of <u>excitement</u> — cold fusion would make it possible to generate lots of electricity, easily and cheaply.

5) After the press conference, the experiments and data were <u>shared</u> with other scientists so they could <u>repeat</u> the experiments. But <u>few</u> managed to reproduce the results <u>reliably</u> — so it hasn't been accepted as a <u>realistic</u> method of energy production.

Fusion and fission — don't get them mixed up

Nuclear fission and nuclear fusion may sound very similar but they're very different processes.
Don't get them mixed up. Just remember, in nuclear <u>fusion</u> you're <u>fusing</u> two nuclei together.
In nuclear fission you're splitting a larger nucleus into two smaller nuclei.

The Life Cycle of Stars

Stars go through <u>many traumatic stages</u> in their lives — just like teenagers.

Protostar

1) Stars <u>initially form</u> from <u>clouds of DUST AND GAS</u>. The <u>force of gravity</u> makes the gas and dust <u>spiral in together</u> to form a <u>protostar</u>.

2) <u>Gravitational energy</u> is converted into <u>heat energy</u>, so the <u>temperature rises</u>. When the temperature gets <u>high enough</u>, hydrogen nuclei undergo <u>nuclear fusion</u> to form <u>helium nuclei</u> and give out massive amounts of <u>heat and light</u>. A star is born. Smaller masses of gas and dust may also pull together to make <u>planets</u> that orbit the star.

Main Sequence Star

3) The star immediately enters a <u>long stable period</u>, where the <u>heat created</u> by the nuclear fusion provides an <u>outward pressure</u> to <u>balance</u> the <u>force of gravity</u> pulling everything <u>inwards</u>. The star maintains its energy output for <u>millions of years</u> due to the <u>massive amounts of hydrogen</u> it consumes. In this <u>stable</u> period it's called a <u>MAIN SEQUENCE STAR</u> and it lasts <u>several billion years</u>. (The Sun is in the middle of this stable period — or to put it another way, the <u>Earth</u> has already had <u>half its innings</u> before the Sun <u>engulfs</u> it!)

Stars much bigger than the Sun

Stars about the same size as the Sun

4) Eventually the <u>hydrogen</u> begins to <u>run out</u>. Heavier elements such as iron are made by nuclear fusion of <u>helium</u>. The star then <u>swells</u> into a <u>RED GIANT</u>, if it's a small star, or a <u>RED SUPER GIANT</u> if it's a big star. It becomes <u>red</u> because the surface <u>cools</u>.

Red Giant ➡️ **White Dwarf**

Red Super Giant

5) A <u>small-to-medium</u>-sized star like the Sun then becomes unstable and <u>ejects</u> its <u>outer layer</u> of <u>dust and gas</u> as a <u>PLANETARY NEBULA</u>.

6) This leaves behind a hot, dense solid core — a <u>WHITE DWARF</u>, which just cools down to a <u>BLACK DWARF</u> and eventually disappears.

Neutron Star... ➡️ **...or Black Hole**

Supernova

7) <u>Big stars</u>, however, start to <u>glow brightly again</u> as they undergo more <u>fusion</u> and <u>expand and contract several times</u>, forming elements as <u>heavy as iron</u> in various <u>nuclear reactions</u>. Eventually they <u>explode</u> in a <u>SUPERNOVA</u>, forming elements <u>heavier than iron</u> and ejecting them into the universe to <u>form new planets and stars</u>.

8) The <u>exploding supernova</u> throws the outer layers of <u>dust and gas</u> into space, leaving a <u>very dense core</u> called a <u>NEUTRON STAR</u>. If the star is <u>big enough</u> this will become a <u>BLACK HOLE</u>.

Only big stars become black holes

The early universe contained <u>only hydrogen</u>, the simplest and lightest element. It's only thanks to nuclear fusion inside stars that we have any of the other <u>naturally occurring elements</u>. Remember — the heaviest element produced in stable stars is iron, but it takes a <u>supernova</u> (or a lab) to create <u>the rest</u>.

Warm-Up and Exam Questions

The end of the last section — they just go far too quickly. Make sure you've understood it all by doing these questions (and the revision summary) before you whizz off into the sunset.

Warm-Up Questions

1) Name two elements often used as nuclear fuel.
2) What does nuclear fission produce in addition to energy? Why is this a problem?
3) a) Explain why people were excited about a successful cold fusion experiment.
   b) Explain why the results of this experiment haven't yet been accepted by the scientific community.
4) What are stars formed from?
5) Will our Sun become a black hole? Explain your answer.
6) At the end of its main sequence phase, what does a small star become?

Exam Questions

1   Nuclear power is a useful way of generating electricity but it also generates nuclear waste.

(a)   Most of the waste produced by a nuclear power station is described as 'low level' nuclear waste.

(i)   Explain what is meant by the term 'low level'.

(1 mark)

(ii)   Describe how 'low level' nuclear waste is disposed of.

(1 mark)

(b)   One way that you can dispose of intermediate and high level nuclear waste is by burying it deep underground. Explain why nuclear waste can only be buried in places that are geologically stable.

(2 marks)

2   The table shows some information about various elements and isotopes.

Element/isotope	Deuterium	Hydrogen	Krypton	Plutonium	Thorium	Tin
Relative mass	2	1	84	239	232	119

(a)   Name the **two** substances in the table that would be most likely to be used in a fusion reaction.

(2 marks)

(b)   Explain why scientists are interested in developing fusion power.

(2 marks)

(c)   Explain why fusion is not used to generate electricity at present.

(1 mark)

Exam Questions

3   Stars go through many stages in their lives.

(a)   Describe how a star is formed.

(3 marks)

(b)   The stable period of a main sequence star can last millions of years.
Explain why main sequence stars undergo a stable period.

(1 mark)

(c)   When main sequence stars begin to run out of hydrogen in their core, they swell
and become either a red giant or a super red giant depending on their size.

(i)   Describe what happens to small stars after their red giant phase.

(3 marks)

(ii)   Describe what happens to big stars after their super red giant phase.

(4 marks)

4   Nuclear power stations use nuclear reactors and nuclear fuel to produce electricity.

(a)   Complete the sentences below on nuclear power stations.

(i)   Inside a nuclear reactor, nuclei split up and release .... energy.

(1 mark)

(ii)   This energy is used to produce .... which is used

to drive a turbine connected to an electricity generator.

(1 mark)

(b)   The diagram below shows a chain reaction that takes place in a nuclear reactor.

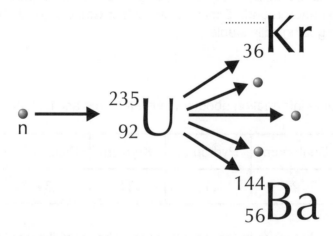

(i)   Complete the diagram to show the mass number of the krypton (Kr) produced.
Show how you worked out your answer.

(2 marks)

(ii)   Use the diagram above to describe how a chain reaction
is set up in a nuclear reactor.

(4 marks)

Revision Summary for Section 9

Phew... what a relief — you made it to the end of the book. But don't run off to put the kettle on just yet — make sure that you really know your stuff with these revision questions.

1) Explain how the experiments of Rutherford and Marsden led to the nuclear model of the atom.

2) Draw a table stating the relative mass and charge of the three basic subatomic particles.

3) Explain what isotopes are. Give an example.
Do stable or unstable isotopes undergo nuclear decay?

4) True or false: radioactive decay can be triggered by certain chemical reactions.

5) Give three sources of background radiation.

6) What are the units of radiation dose?

7) List two places where the level of background radiation is higher than normal and explain why.

8) Name three occupations that have an increased risk of exposure to radiation.

9) What type of subatomic particle is a beta particle?

10)*Complete the following nuclear equations by working out the missing numbers shown
by the dotted lines: a) $^{131}_{53}I \rightarrow \ ^{...}_{...}Xe + \ ^{0}_{-1}e$ b) $^{...}_{...}Gd \rightarrow \ ^{144}_{62}Sm + \ ^{4}_{2}He$

11) Sketch the paths of an alpha particle and a beta particle travelling through an electric field.

12)*The activity of a radioactive sample is 840 Bq. Four hours later it has fallen to 105 Bq.
Find the half-life of the sample.

13) Sketch a typical graph of activity against time for a radioactive source.
Show how you can find the half-life from your graph.

14) What substances could be used to block:
a) alpha radiation, b) beta radiation, c) gamma radiation?

15) Describe in detail how radioactive sources are used in each of the following:
a) treating cancer,   b) medical tracers,   c) sterilising food,   d) thickness gauges.

16) What is the main environmental problem associated with nuclear power?

17) What is nuclear fusion? Why is it difficult to construct a working fusion reactor?

18) Briefly explain why cold fusion isn't accepted as a realistic method of energy production.

19) What is a supernova?

*Answers on page 281.

Section 1

Page 19

Warm-Up Questions

1) Plant cells have a rigid cell wall whereas animal cells don't have a cell wall. Plant cells have a vacuole whereas animal cells don't have a vacuole. Plant cells contain chloroplasts whereas animal cells don't.

2) a nucleus

3) Any two of, e.g. cell membrane / cytoplasm / cell wall / genetic material.

4) (a) To carry oxygen.

   (b) Any two of, e.g. concave shape gives the red blood cells a large surface area for absorbing oxygen / concave shape helps the red blood cells pass through capillaries to body cells / no nucleus maximises the space for haemoglobin.

5) (a) To get the male DNA to the female DNA.

   (b) It has a long tail/streamlined head to help it swim to the egg. / It has a lot of mitochondria to provide energy. / It carries enzymes in its head to digest the egg cell membrane.

6) An organ system is a group of organs working together to perform a particular function.

Exam Questions

1 (a) C *(1 mark)*

  (b) chlorophyll *(1 mark)*

  (c) E.g. the palisade cell has a tall shape/a large surface area for absorbing carbon dioxide / palisade cells have a thin shape, so lots of them can be packed together at the top of a leaf *(1 mark)*.

2 (a) leaves *(1 mark)*, plump/turgid *(1 mark)*, stomata/pores *(1 mark)*, gases *(1 mark)*, photosynthesis *(1 mark)*.

  (b) They become flaccid *(1 mark)*.

  (c) The guard cells go flaccid to close the stomata *(1 mark)*. This saves the plant water without losing out on photosynthesis *(1 mark)*.

Pages 26-27

Warm-Up Questions

1) A catalyst is a substance that increases the speed of a reaction, without being changed or used up in the reaction.

2) The optimum pH is the pH at which the enzyme works best.

3) They break down big molecules into smaller ones.

4) (a) protease/pepsin

   (b) lipase

   Proteases break down proteins, and lipases break down lipids (fats).

5) (a) sugars/maltose

   (b) amino acids

   (c) glycerol and fatty acids

6) Proteases are used to pre-digest the protein in some baby foods so it's easier for a baby to digest.

Exam Questions

1 (a) The enzyme has a specific shape which will only fit with one substrate *(1 mark)*.

  (b) In the wrong conditions (e.g. high temperatures), the bonds in the enzyme are broken/the enzyme changes shape, so the substrate can no longer fit into it/the enzyme won't work anymore *(1 mark)*.

  If you heat a substance you supply it with energy and it moves about more. This helps things to react faster. But if you heat an enzyme too much, it jiggles about such a lot that it ends up breaking some of the bonds that hold it together and it loses its shape. A similar thing happens with pH — the wrong pH disrupts the bonds and the shape is changed.

2 (a)

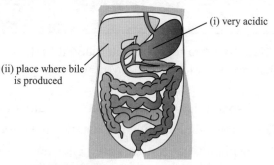

(i) very acidic

(ii) place where bile is produced

(1 mark for each)

The acidic part is the stomach.
The liver is where bile is produced.

  (b) (i) It stores bile until it is ready to be released *(1 mark)*.

      (ii) It produces protease, amylase and lipase enzymes *(1 mark)* and releases them into the small intestine *(1 mark)*.

3 (a) isomerase *(1 mark)*

  (b) Fructose is sweeter than glucose *(1 mark)*. This means the food industry can use less of it in foods, which is good for making slimming foods *(1 mark)*.

4 (a) Accept answers between 38 °C and 40 °C *(1 mark)*.

  (b) Enzyme B, because it has an unusually high optimum temperature which it would need to work in the hot vent *(1 mark)*.

  (c) It would break down the proteins in stains such as blood and grass on clothing *(1 mark)*. It would not be denatured by high-temperature washes *(1 mark)*.

5 (a) amylase/carbohydrase *(1 mark)*

  (b) Any three of, e.g. temperature / volume of solution / concentration of starch solution / concentration of enzyme solution *(1 mark each)*.

  (c) At pH 8 *(1 mark)*, because this is the pH in the part of the gut where this enzyme works/because most enzymes work best at around a neutral pH *(1 mark)*.

6 How to grade your answer:

   0 marks: No advantages or disadvantages of using enzymes in industry are given.

   1-2 marks: There is a brief description of one advantage and one disadvantage of using enzymes in industry.

   3-4 marks: At least two advantages and two disadvantages of using enzymes in industry are given. The answer has a logical structure and spelling, grammar and punctuation are mostly correct.

   5-6 marks: The answer gives at least three advantages and three disadvantages of using enzymes in industry. The answer has a logical structure and uses correct spelling, grammar and punctuation.

Here are some points your answer may include:

Advantages:

Enzymes are specific, so they only catalyse the reaction you want them to.

Enzymes mean that you can use lower temperatures and pressures, which means a lower cost as it saves energy.

Enzymes work for a long time, so after the initial cost of buying them, they can be continually used.

Enzymes are biodegradable and therefore cause less environmental pollution.

Disadvantages:

Some people can develop allergies to the enzymes (e.g. in biological washing powders).

Enzymes can be denatured by even a small increase in temperature. They're also susceptible to poisons and changes in pH, which means that the conditions in which they work must be tightly controlled.

Enzymes can be expensive to produce.

Contamination of the enzyme with other substances can affect the reaction.

Page 34

Warm-Up Questions

1) Respiration is the process of breaking down glucose to release energy, which happens in every cell.

2) glycogen

3) Diffusion is the spreading out of particles from an area of high concentration to an area of low concentration.

4) It is a plant cell with a shrunken cytoplasm and a cell membrane that has pulled away from the cell wall.

Exam Questions

1 (a) B — it has a higher concentration of glucose outside the cell than inside *(1 mark)*, so glucose will diffuse into the cell *(1 mark)*.

   (b) E.g. water *(1 mark)*

2 (a) Because during exercise the muscles need more energy from respiration *(1 mark)*, and this respiration requires oxygen *(1 mark)*.

   (b) Because there is an oxygen debt / oxygen is needed to break down the lactic acid that has built up *(1 mark)*.

Page 38

Warm-Up Questions

1) artery, vein, capillary

2) one

3) C — pulmonary vein and vena cava, atria, ventricles, pulmonary artery and aorta

Exam Questions

1 (a) (i) aorta, vena cava, pulmonary artery, pulmonary vein *(1 mark each)*.

      (ii) (right) atrium *(1 mark)*

   (b) (i) four *(1 mark)*

      (ii) valves *(1 mark)*

      (iii) Valves prevent the blood flowing backwards/in the wrong direction *(1 mark)*.

2 A — 4 *(1 mark)*

   B — 1 *(1 mark)*

   C — 3 *(1 mark)*

   D — 2 *(1 mark)*

Section 2

Page 45

Warm-Up Questions

1) deoxyribonucleic acid

2) The bases are A (adenine), T (thymine), G (guanine) and C (cytosine). A pairs with T and G pairs with C.

3) The detergent breaks down the cell membranes. The salt makes the DNA stick together.

4) A gene is a section of DNA. It contains the instructions to make a specific protein.

5) E.g. a mutation in a gene on a bacterial plasmid that makes it resistant to an antibiotic.

Exam Questions

1 (a) The results suggest that Mr X was at the crime scene, because his DNA profile has the same pattern as the DNA profile from the blood at the crime scene *(1 mark)*.

   (b) It is true that usually everyone's DNA is unique *(1 mark)*, but identical twins have the same DNA and so would have identical genetic fingerprints *(1 mark)*.

2 (a) G—G—C—A—A—A—C—C—C *(1 mark)*.

   (b) The DNA double helix first 'unzips' to form two single strands *(1 mark)*. The bases on free-floating nucleotides pair up with the complementary bases on the DNA *(1 mark)*. Cross links form between the bases and the old DNA strands, and the nucleotides are joined together to form double strands *(1 mark)*.

   (c) They used X-ray data which showed that DNA has a helical structure *(1 mark)*. Other data they used showed that the amount of A and G matched the amount of T and C *(1 mark)*.

Page 53

Warm-Up Questions

1) When a cell reproduces itself by splitting to form two identical offspring.

2) one parent

3) mitosis

4) E.g. sickle cell anaemia.

Exam Questions

1 (a) They are undifferentiated cells *(1 mark)* that can develop into different types of/specialised cells *(1 mark)*.

   (b) They could be grown into a particular type of cell, which could then be used to replace faulty cells *(1 mark)*.

   (c) How to grade your answer:

     0 marks: No reasons for or against using embryos to create stem cells are given.

     1-2 marks: There is a brief description of one reason for or against using embryos to create stem cells for research.

     3-4 marks: There is a description of more than one reason for or against using embryos to create stem cells for research. The answer has a logical structure and spelling, grammar and punctuation are mostly correct.

     5-6 marks: There is a detailed and balanced description of the reasons for and against using embryos to create stem cells for research. The answer has a logical structure and uses correct spelling, grammar and punctuation.

     Here are some points your answer may include:

     Reasons for:

     Some people believe that curing patients who already exist and are suffering is more important than the rights of embryos.

     The embryos used in stem cell research are usually unwanted ones that would probably be destroyed if they weren't used for research.

     Reasons against:

     Some people feel that human embryos shouldn't be used for experiments since each one is a potential human life.

     Some people think scientists should concentrate on finding and developing other sources of stem cells, so people could be helped without having to use embryos.

2 (a) sperm and ova *(1 mark)*

   (b) 23 *(1 mark)*

   (c) 46 *(1 mark)*

   (d) half *(1 mark)*

3 (a) (i) Each cell should contain only three chromosome arms *(1 mark)*, and there should be one of each type *(1 mark)* as shown:

Remember that there are two divisions in meiosis. In the first division, one chromosome from each pair goes into each of two new cells. In the second division, both those cells divide again, with one half of each chromosome going into each of the new cells.

      (ii) They contain half the genetic material that the original cell contained *(1 mark)*.

(b) Any three of, e.g. it involves two divisions, instead of one / it halves the chromosome number, rather than keeping it constant / it produces genetically different cells, not genetically identical cells / it produces sex cells/gametes, not body cells *(1 mark for each)*.

Pages 62-63

Warm-Up Questions

1) X and Y chromosomes

2) They're different versions of the same gene.

3) The allele which causes cystic fibrosis is a recessive allele, so people can have one copy of the allele and not have the disorder/show any symptoms.

4) It's a genetic disorder where a baby is born with extra fingers or toes.

5) A process where embryos are fertilised in a laboratory and then implanted into the mother's womb.

6) 1:1

Exam Questions

1 (a) No, because polydactyly isn't a significant health issue / embryos are only screened for serious genetic disorders *(1 mark)*.

(b) Certain types of gene increase the risk of cancer, but they aren't a definite indication that the person will develop cancer *(1 mark)*.
Also some types of cancer can be treated successfully *(1 mark)*.
It can be argued that it isn't right to destroy an embryo because it might develop a disease which could be treatable *(1 mark)*.

(c) E.g. it implies that people with genetic problems are undesirable. / The rejected embryos (which could have developed into humans) are destroyed. / They may be worried that it could lead to embryo screening being allowed for other traits. *(1 mark)*

Extracts can be a bit scary — all that scientific information in a big wodge. Try reading the extract once, then reading the questions and then reading the extract again, underlining any useful bits. The extract's there to help you with the questions — so use it.

2 (a) White flowers *(1 mark)*

(b) (i) FF *(1 mark)*

(ii) ff *(1 mark)*

(iii) Ff *(1 mark)*

3 (a)

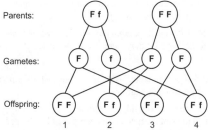

Parents: F f   F F
Gametes: F   f   F   F
Offspring: F F   F f   F F   F f
            1   2   3   4

(1 mark for gametes correct, 1 mark for offspring correct)

(b) They will all be unaffected *(1 mark)*.

(c) (i) 1 in 2 / 50% *(1 mark)*

(ii) 2 and 4 *(1 mark)*

Pages 69-70

Warm-Up Questions

1) Moving genes (sections of DNA) from one organism to another so that it produces useful biological products.

2) E.g. insulin.

3) A clone is an organism that is genetically identical to another organism.

4) With an electric shock.

5) A reduced gene pool is where there are fewer different alleles in a population.

6) Organisms with the best characteristics are selected and bred with each other. The best of their offspring are then selected and bred. This process is repeated over several generations.

Exam Questions

1 (a) E.g. to cut the gene out of the donor organism's DNA *(1 mark)*.

(b) E.g. to give resistance to herbicides/frost damage/disease. / Vitamin A production in food crops/rice *(1 mark)*.

(c) Any three of, e.g. inserted genes might have unexpected harmful effects. / Some people worry that the engineered DNA might 'escape', e.g. weeds could gain rogue genes from a crop that's had genes for herbicide resistance inserted into it. / Some people think it's wrong to genetically engineer other organisms purely for human benefit. / People worry that we won't stop at engineering plants and animals, and that in the future people may be able to decide the characteristics they want their children to have. / The evolutionary consequences of genetic engineering are unknown, so some people think it's irresponsible to carry on when we're not sure what the impact on future generations might be *(1 mark for each)*.

Genetic engineering is a really controversial issue — and it could easily come up in the exam. Make sure you know some of its pros and cons.

2 (a) Remove the genetic material/nucleus from the unfertilised egg cell of a sheep *(1 mark)*. Insert a complete set of chromosomes from an adult sheep's body cell into the empty egg cell *(1 mark)*. Stimulate the egg cell with an electric shock to make it divide *(1 mark)*. When the embryo is a ball of cells, implant it into a female sheep/surrogate mother to develop *(1 mark)*.

(b) Female. Dolly is a clone, so she must have exactly the same genetic material as her parent *(1 mark)*.

3 (a) So that it does not contain Beatrix's genes / So it can receive the genetic information from Brenda *(1 mark)*.

(b) So that the offspring has the same genetic material as Brenda/is a clone of Brenda *(1 mark)*.

(c) Brenda, because it is her genetic information that is being transmitted to the baby *(1 mark)*.

Belinda gave birth to the baby mouse, but the egg cell that was implanted into her uterus came from Beatrix, and the nucleus (where all of the genetic material came from) was from Brenda. The baby mouse is therefore a clone of Brenda.

4 (a) By using selective breeding *(1 mark)* — only the fastest horses are chosen for breeding *(1 mark)*, and so over time this characteristic becomes predominant in the offspring *(1 mark)*.

(b) E.g. it reduces the gene pool. / The horses are inbred, and so may be more prone to genetic disorders. / An inbred population of horses may all be susceptible to the same diseases *(1 mark)*.

5 (a) E.g. cloning organs for transplant / cloning adult body cells to produce human embryos which could supply stem cells for stem cell therapy *(1 mark for each)*.

(b) How to grade your answer:

0 marks:	No potential disadvantages given.
1-2 marks:	Brief description of one or two potential disadvantages of cloning.
3-4 marks:	Some discussion of at least three potential disadvantages of cloning. The answer has a logical structure and spelling, punctuation and grammar are mostly correct.
5-6 marks:	A clear, detailed and full discussion of at least five potential disadvantages of cloning. The answer has a logical structure and uses correct spelling, grammar and punctuation.

Here are some points your answer may include:

Cloning results in a reduced gene pool, which may leave a cloned population more susceptible to disease.

It's possible that cloned animals might not be as healthy as normal ones.

The cloning process often fails, e.g. it took hundreds of attempts to clone Dolly the sheep.

Clones are often born with genetic defects.

Cloned mammals' immune systems are sometimes unhealthy, so they suffer from more diseases.

Cloned mammals might not live as long.

Page 74

Warm-Up Questions

1) Fossils are the remains of organisms from many years ago, which are found in rocks.

2) E.g. an animal's burrow / a plant's roots / a footprint.

3) Some fossils are destroyed by geological activity, e.g. the movement of tectonic plates may have crushed fossils already formed in the rocks.

4) A species that doesn't exist anymore.

5) Isolation is where populations of a species are separated.

Exam Questions

1 (a) Inside the amber there is no oxygen or moisture *(1 mark)* so the insect remains can't decay / so decay microbes can't survive *(1 mark)*.

(b) E.g. a volcanic eruption / a collision with an asteroid *(1 mark)*.

2 (a) Teeth don't decay easily so they can last a long time when buried *(1 mark)*. They're eventually replaced by minerals as they decay (leaving a rock-like substance in the shape of the teeth) *(1 mark)*.

(b) Parts of Stegosaurus that were made of soft tissue will have decayed away completely, without being fossilised *(1 mark)*.

3 E.g. two populations of the original bird species became isolated/separated by a physical barrier/water *(1 mark)*. Conditions on the two islands were slightly different *(1 mark)*. Each population showed genetic variation because they had a wide range of alleles *(1 mark)*. Different characteristics became more common in each population due to natural selection *(1 mark)*. Over time the two populations became so different that they couldn't interbreed any more *(1 mark)*.

Page 75

Revision Summary for Section 2

18) BB and bb

Section 3

Pages 81-82

Warm-Up Questions

1) Carbon dioxide, water, (sun)light, chlorophyll.

2) A limiting factor is something that stops photosynthesis from happening any faster.

3)

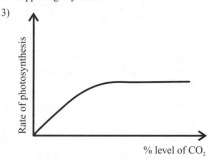

4) E.g. by using a paraffin heater.

5) respiration

Exam Questions

1 Making cell walls — cellulose *(1 mark)*
Making proteins — amino acids *(1 mark)*
Storing energy — starch *(1 mark)*

2 (a) epidermis/epidermal tissue *(1 mark)*

(b) Any two of, e.g. mesophyll tissue / xylem / phloem *(1 mark each)*

3 (a) (i) The enzymes needed for photosynthesis work more slowly at low temperatures, so the rate of photosynthesis is slower at 10 °C than at 20 °C *(1 mark)*.

(ii) Photosynthesis is happening quickly at 40 °C but at 50 °C it has stopped. At 50 °C the enzymes are denatured/damaged/the plant dies *(1 mark)*.

(b) In the experiment the rate was highest at this temperature *(1 mark)*, but the optimum could actually be anywhere between 30 °C and 50 °C (where no measurements were made) *(1 mark)*.

4 (a) By counting the number of bubbles produced/measuring the volume of gas produced, in a given time/at regular intervals *(1 mark)*.

(b) (i) The rate of photosynthesis/number of bubbles/volume of gas *(1 mark)*.

(ii) The light intensity *(1 mark)*.

(c) E.g. carbon dioxide concentration in the water/temperature/the plant being used *(1 mark)*.

5 (a) A label anywhere on the sloping part of the graph, before it levels off *(1 mark)*.

(b) Carbon dioxide concentration / temperature / amount of chlorophyll / amount of water *(1 mark)*.

6 (a) Chlorophyll *(1 mark)*.

(b) (i)

(1 mark)

(ii) Plants need both chlorophyll and light to photosynthesise and produce glucose which is stored as starch *(1 mark)*. There is only chlorophyll in the green area of the plant *(1 mark)*, and light can only reach parts of the leaf not covered by black paper *(1 mark)*.

Page 88

Warm-Up Questions

1) The constant stream of water through a plant.

2) In areas of a plant that are growing, e.g. the roots and shoots.

3) It contains plant hormones, which make the cuttings produce roots rapidly and start growing as new plants.

4) The growth of a plant in response to light.

5) auxin

Exam Questions

1 (a) It has a hair-like shape that sticks out into the soil, creating a large surface area for absorbing water *(1 mark)*.

(b) The concentration of mineral ions is higher inside the root hair cell than in the soil around it *(1 mark)*. So mineral ions are absorbed by active transport *(1 mark)* against a concentration gradient/using energy from respiration *(1 mark)*.

2 (a) The right side *(1 mark)*.

(b) Auxin is produced in the tip and moves backwards/down the stem *(1 mark)*, so there is more auxin on the right side *(1 mark)*. This makes the cells grow/elongate faster on the right side, so the stem grows to the left *(1 mark)*.

(c) Because light would change the distribution of the auxin *(1 mark)*.

Page 92

Warm-Up Questions

1) Where an organism is found.

2) A quadrat is a square frame enclosing a known area.

3) The median is the middle value, in order of size, so it's 8.

4) transect

Exam Questions

1 (a) 21 *(1 mark)*

   (b) (i) 6 + 15 + 9 + 14 + 20 + 5 + 3 + 11 + 10 + 7 = 100 *(1 mark)*

   100/10 = 10 dandelions per m² *(1 mark)*

   (ii) 90 × 120 = 10 800 m² *(1 mark)*

   10 × 10 800 = 108 000 dandelions in field F *(1 mark)*

   (c) Anna's, because she used a larger sample size (20 quadrats per field, whereas Paul only used 10) *(1 mark)*.

Section 4

Page 99

Warm-Up Questions

1) Mass number is the total number of protons and neutrons in an atom. Atomic number is the number of protons in an atom.

2) The groups of three Döbereiner organised the elements into, based on their chemical properties.

3) E.g. the number of electrons in the outer shell of an atom of the element.

4) periods

Exam Questions

1 (a) Isotopes are different atomic forms of the same element *(1 mark)*, which have the same number of protons *(1 mark)* but a different number of neutrons *(1 mark)*.

   (b)

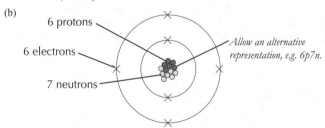

6 protons
6 electrons
7 neutrons

Allow an alternative representation, e.g. 6p7n.

   (1 mark each for the correct number and placement of protons, neutrons and electrons)

2 (a) 1 *(1 mark)*

   (b) The relative mass of electrons is very small *(1 mark)*.

3 D *(1 mark)*

4 (a) Newlands listed elements in rows of seven in order of their relative atomic mass *(1 mark)*.

   (b) E.g. Newlands' groups contained elements that didn't have similar properties, e.g. carbon and titanium *(1 mark)*. He mixed up metals and non-metals, e.g. oxygen and iron *(1 mark)*. He didn't leave any gaps for elements that hadn't been discovered yet unlike Mendeleev *(1 mark)*. This meant that Newlands' Octaves broke down on the third row *(1 mark)*.

Page 104

Warm-Up Questions

1) it increases

2) hydrogen

3) E.g. sodium bromide, potassium chloride (any Group 1 halide).

You could have written your answer as a formula too, e.g. KCl.

4) $2Fe_{(s)} + 3Cl_{2(g)} \rightarrow 2FeCl_{3(s)}$

5) $Cl_{2(g)} + 2KBr_{(aq)} \rightarrow Br_{2(aq)} + 2KCl_{(aq)}$

Exam Questions

1 (a) Chlorine — gas *(1 mark)*

   Bromine — liquid *(1 mark)*

   Iodine — solid *(1 mark)*

   (b) Arrow should be pointing upwards. *(1 mark)*

   (c) (i) displacement *(1 mark)*

   Chlorine is displacing iodine.

   (ii) iodine/I/I₂ *(1 mark)*

2 (a) They have a single outer electron which is easily lost so they are very reactive *(1 mark)*.

   (b) As you go down Group 1, the outer electron is further from the nucleus *(1 mark)*. This means that less energy is needed to remove the outer electron and so the alkali metal is more reactive *(1 mark)*.

   (c) hydrogen *(1 mark)* and a metal hydroxide *(1 mark)*

   (d) alkaline *(1 mark)*

Pages 110-111

Warm-Up Questions

1) A high boiling point.

2) When ionic compounds are dissolved the ions separate and are free to move in the solution. These free-moving charged particles allow the solution to carry electric current.

3) positive ions

4) negative ions

5) $Al(OH)_3$

Exam Questions

1 (a) Sodium will lose the electron from its outer shell to form a positive ion *(1 mark)*. Fluorine will gain an electron to form a negative ion *(1 mark)*.

   (b) Ionic compounds have a giant ionic lattice structure *(1 mark)*. The ions form a closely packed regular lattice arrangement *(1 mark)*, held together by strong electrostatic forces of attraction between oppositely charged ions *(1 mark)*.

2 (a) $MgCO_3$ *(1 mark)*

   (b) Li_2SO_4 *(1 mark)*

3 (a) lithium oxide *(1 mark)*

   (b) (i) and (ii)

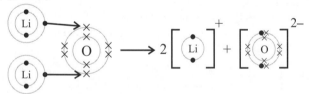

(1 mark for arrows shown correctly, 1 mark for correct electron arrangement and charge on lithium ion, 1 mark for correct electron arrangement and charge on oxygen ion).

4 (a)

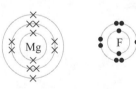

   (1 mark)   *(1 mark)*

   (b) Mg^{2+} *(1 mark)* and F^- *(1 mark)*

   (c) MgF_2 *(1 mark)*

   (d) There are electrostatic forces of attraction between the ions *(1 mark)*.

   (e) (i) There are strong electrostatic forces between the ions *(1 mark)* so a large amount of energy is needed to break these bonds/overcome these forces *(1 mark)*.

   (ii) When the magnesium fluoride is molten the ions can move about and carry charge/a current through the liquid *(1 mark)*.

5 (a)

	Potassium atom, K	Potassium ion, K⁺	Chlorine atom, Cl	Chloride ion, Cl⁻
Number of electrons	19	**18**	**17**	**18**

(2 marks for all three columns correct, otherwise 1 mark for any two columns correct)

(b)

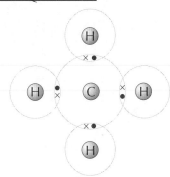

(2 marks — 1 mark for correct electron arrangements, 1 mark for correct arrow and charges on ions)

Page 117

Example

A is simple molecular, B is giant metallic,
C is giant covalent, D is giant ionic

Pages 118-119

Warm-Up Questions

1) It's where atoms share electrons.

2) Because the intermolecular forces between the chlorine molecules are very weak.

3) E.g. diamond is very hard and graphite is fairly soft.
   Graphite conducts electricity and diamond doesn't.

4) E.g. silicon dioxide/silica.

Exam Questions

1

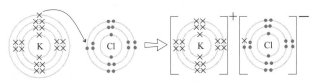

(1 mark for bonds shown correctly, 1 mark for correct number of atoms shown)

2 (a) (i)   giant covalent *(1 mark)*

    (ii) giant covalent *(1 mark)*

    (iii) simple molecular *(1 mark)*

(b) It doesn't contain any ions or free electrons to carry the charge *(1 mark)*.

(c) Each carbon atom has a delocalised electron that's able to carry the charge *(1 mark)*.

(d) All of the atoms in silicon dioxide and in graphite are held together by strong covalent bonds *(1 mark)*. In bromine the atoms within each molecule are held together with strong covalent bonds but the forces between these molecules are weak *(1 mark)*.

In order to melt, a substance has to overcome the forces holding its particles tightly together in the rigid structure of a solid. If the forces between the particles are weak, this is easy to do and doesn't take much energy at all. But if the forces are really strong, like in a giant covalent structure, you have to provide loads of heat to give the particles enough energy to break free.

3 (a) E.g:

(1 mark for showing two different sizes of atoms, 1 mark for showing irregular arrangement)

(b) The regular arrangement of atoms in iron means that they can slide over each other meaning iron can be bent *(1 mark)*. Steel contains different sized atoms which distorts the layers of iron atoms *(1 mark)* making it more difficult for them to slide over each other *(1 mark)*.

4 (a) solid *(1 mark)*

(b) (i)   giant covalent *(1 mark)*

    (ii) giant ionic *(1 mark)*

(c)

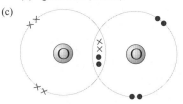

(1 mark for correct number of electrons in each shell, 1 mark for showing covalent bond correctly)

(d) (i)   It has delocalised electrons *(1 mark)* which are free to move through the whole structure and carry an electrical charge *(1 mark)*.

    (ii) yes *(1 mark)*

Page 124

Warm-Up Questions

1) A shape memory alloy (about half nickel and half titanium) / a smart material / a smart material that returns to a 'remembered' shape when it's heated above a certain temperature.

2) The higher the melting point the stronger the forces holding the polymer chains together.

3) crosslinks

4) E.g. the starting materials and the reaction conditions.

Exam Questions

1 (a) (i)   nm *(1 mark)*

    (ii) surface area *(1 mark)*, volume *(1 mark)*

(b) (i)   E.g. the CNTs provide strength *(1 mark)*.

    (ii) fullerenes *(1 mark)*

(c) E.g. in computer chips / in sensors / as catalysts / delivering drugs / in cosmetics / in lubricants *(1 mark)*.

2 (a) B *(1 mark)*

(b) C (accept A) *(1 mark)*

(c) A *(1 mark)*

Page 126

Top Tip

a) 30.0%

b) 88.9%

c) 48.0%

d) 65.3%

Page 127

Top Tip

CH_4

Page 129

Warm-Up Questions

1) The relative atomic mass.

2) The relative formula mass.

3) A mole is the relative formula mass of a substance, in grams.

4) One mole of O_2 weighs $16 \times 2 = 32$ g.

Exam Questions

1 (a) (i) Relative atomic mass *(1 mark)*.

    (ii) Boron-11 has one more neutron in its nucleus than boron-10 *(1 mark)*.

    (iii) Boron has two isotopes, and the relative atomic mass is an average value of both isotopes, taking into account how much there is of each isotope *(1 mark)*.

  (b) (i) M_r of $BF_3 = 11 + (19 \times 3) = 68$ *(1 mark)*

    (ii) M_r of $B(OH)_3 = 11 + (17 \times 3) = 62$ *(1 mark)*

2 (a) $100 - 60 = 40\%$ *(1 mark)*

  (b) 40 g of sulfur combine with 60 g of oxygen.

    S = 40   O = 60

    $40 \div 32$ $60 \div 16$

    $= 1.25$ $= 3.75$

    $1.25 \div 1.25 = 1$ $3.75 \div 1.25 = 3$

    Therefore, the formula of the oxide is SO_3 *(2 marks, otherwise 1 mark for correct working)*

Any way of getting to the right answer is fine, as long as you show your working. For example, you might have got a ratio of 125 : 375 and then divided by 125.

3 (a) 100g .... reacts to give ...   56 g

    1 g .... reacts to give ... $56 \div 100 = 0.56$ g

    2 g .... reacts to give ... $0.56 \times 2 = 1.12$ g   *(1 mark)*

  (b) E.g. When transferring the $CaCO_3$ from the weighing apparatus to the test tube, or the CaO from the test tube to the weighing apparatus some of the solid may be left behind *(1 mark)*.

Page 134

Warm-Up Questions

1) Waste by-products decrease the atom economy of a reaction.

2) 100%.

   All reactions with one product will have 100% atom economy.

3) Because it uses up resources very quickly and produces a lot of waste.

4) Percentage yield $= (4 \div 5) \times 100$

                $= 80\%$

5) Because a low product yield means that resources are wasted rather than being saved for future generations.

Exam Questions

1 M_r of water $= (1 \times 2) + 16 = 18$

  M_r of ethene $= (12 \times 2) + 4 = 28$ *(1 mark)*

  Atom economy $= 28 \div (28 + 18) \times 100$ *(1 mark)*

                $= 61\%$ *(1 mark)*

2 (a) From the equation, 4 moles of CuO $\rightarrow$ 4 moles of Cu

    so 1 mole CuO $\rightarrow$ 1 mole Cu *(1 mark)*

    $63.5 + 16 = 79.5$ g CuO $\rightarrow$ 63.5 g Cu *(1 mark)*

    1 g CuO $\rightarrow$ $63.5 \div 79.5 = 0.8$ g Cu

    4 g CuO $\rightarrow$ $0.8 \times 4 = 3.2$ g Cu *(1 mark)*

Percentage yields are always less than 100% — so if you get an answer that's more than 100%, it must be wrong.

  (b) Percentage yield $= (2.8 \div 3.2) \times 100$ *(1 mark)*

                 $= 87.5\%$ *(1 mark)*

3 How to grade your answer:

  0 marks: No reasons why yields are always less than 100% are given.

  1-2 marks: Brief description of one reason why yields are less than 100% is given.

  3-4 marks: Two or three reasons why yields are always less than 100% are given. The answer has a logical structure and spelling, grammar and punctuation are mostly correct.

  5-6 marks: At least four reasons why yields are always less than 100% are given. The answer has a logical structure and uses correct spelling, grammar and punctuation.

Here are some points your answer may include:

Some reactions are reversible. The reactants will never be completely converted to products because the reaction goes both ways.

If a liquid is filtered to remove solid particles, some of the liquid or solid could be lost when it's separated from the reaction mixture.

Transferring solutions from one container to another often leaves behind traces on the containers.

There may be unexpected reactions happening. These reaction will use up reactants, so there's not as much to make the desired product.

4 D *(1 mark)*

Page 135

Revision Summary for Section 4

11) $2Li_{(s)} + Cl_{2(g)} \rightarrow 2LiCl_{(s)}$

16) $CaCl_2$

17)

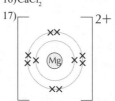

25) a) 40 b) 108 c) 44 d) 84 e) 106 f) 81 g) 56 h) 17

27) a) i) 12.0% ii) 27.3% iii) 75.0%

    b) i) 74.2% ii) 70.0% iii) 52.9%

28) $MgSO_4$

29) 80.3 g

Section 5

Page 141

Warm-Up Questions

1) Blue.

2) Bubble the gas through limewater — the limewater will turn milky if it's carbon dioxide.

3) Instrumental methods are very sensitive, very fast and very accurate.

4) A gas chromatography machine can be attached to a mass spectrometer (GC-MS). The relative molecular mass of a substance can then be read off from the molecular ion peak on the graph the mass spectrometer draws.

Exam Questions

1

Flame colour	Metal ion
green	Cu^{2+}
lilac	K^+
yellow	Na^+
brick-red	Ca^{2+}

(1 mark each)

2 (a) Calcium chloride *(1 mark for calcium, 1 mark for chloride)*.

  (b) Potassium iodide *(1 mark for potassium, 1 mark for iodide)*.

3 (a) 4 *(1 mark)*

(b) 1 *(1 mark)*

(c) 3 *(1 mark)*

Page 145

Warm-Up Questions

1) E.g. bleach / oven cleaner / washing-up liquid.

2) batch production

3) The manufacturer has to prove that the drug meets legal requirements so it works and it's safe.

4) Any two of: e.g. research and development requires the work of lots of highly paid scientists. / The manufacturer has to spend time and money ensuring that the drugs meet legal requirements. / The manufacture of drugs by multi-step batch production is labour-intensive and can't be automated. / The energy and raw materials needed are expensive. / The cost of extraction of raw materials from plants is expensive.

Exam Questions

1 (a) E.g. sulfuric acid *(1 mark)*.

(b) E.g. the start-up costs are huge. / If you run at less than full capacity the process stops being cost-effective *(1 mark)*.

(c) Any two of: e.g. production never stops, so you don't waste time emptying the reactor and setting it up again. / It runs automatically — you only need to interfere if something goes wrong. / The quality of the product is very consistent *(1 mark for each)*.

2 How to grade your answer:

0 marks:	No description of how to extract a substance from a plant is given.
1-2 marks:	One or two of the steps involved in extracting a substance from a plant are given.
3-4 marks:	Three of the steps involved in extracting a substance from a plant are given. The answer has a logical structure and spelling, grammar and punctuation are mostly correct.
5-6 marks:	A detailed explanation of all the steps involved in extracting a substance from a plant is given. The answer has a logical structure and uses correct spelling, grammar and punctuation.

Here are some points your answer may include:

The plant has to be crushed.

The plant has to be boiled and dissolved in a suitable solvent.

The substance can be separated from other chemicals in the plant tissue using chromatography.

In chromatography, spots of different chemicals move up the paper at different speeds, so the chemicals are separated out.

You can extract the chemical you want by cutting the spot out of the chromatography paper and dissolving it off the paper.

3 Sample 3 *(1 mark)*. It has a melting point of 140 °C and it only produces one spot during chromatography, which shows that there are no impurities present *(1 mark)*.

Page 153

Warm-Up Questions

1) E.g. the corrosion of iron is a reaction that happens very slowly. Explosions are very fast reactions.

2) By keeping the milk cool/storing it in a fridge.

3) Any three of, e.g. Increase the temperature (of the acid). / Use smaller pieces of/powdered magnesium. / Increase the acid concentration. / Use a catalyst.

4) Measure the volume of gas given off by collecting it in a gas syringe. / Monitor the mass of a reaction flask from which the gas escapes.

5) It would increase the time taken (i.e. reduce the rate of reaction).

Exam Questions

1 (a) Any two from: e.g. the concentration of sodium thiosulfate/hydrochloric acid / the person judging when the black cross is obscured / the black cross used (size, darkness etc.) *(1 mark each)*.

Judging when a cross is completely obscured is quite subjective — two people might not agree on exactly when it happens. You can try to limit this problem by using the same person each time, but you can't remove the problem completely. The person might have changed their mind slightly by the time they do the next experiment — or be looking at it from a different angle, be a bit more bored, etc.

(b)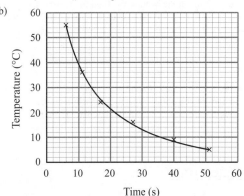

(1 mark for all points plotted correctly, 1 mark for best-fit curve)

(c) As the temperature increases the time decreases, meaning that the reaction is happening faster *(1 mark)*.

(d) Each of the reactions would happen more slowly *(1 mark)*, although they would still vary with temperature in the same way *(1 mark)*.

(e) E.g. by repeating the experiment and taking an average of the results *(1 mark)*.

2 (a) A gas/carbon dioxide is produced and leaves the flask *(1 mark)*.

(b) The same volume and concentration of acid was used each time, with excess marble *(1 mark)*.

(c) E.g. the marble chips were smaller/the temperature was higher *(1 mark)*.

(d) The concentration of the acid is greatest at this point, before it starts being converted into products *(1 mark)*.

Page 159

Warm-Up Questions

1) E.g. by calculating the slope of the line.

2) There are more particles of reactant between the water molecules so collisions between the particles are more likely.

3) They must collide with enough energy.

4) A catalyst is a substance which speeds up a reaction, without being changed or used up in the reaction.

Exam Questions

1 How to grade your answer:

0 marks:	No ways of increasing the rate given.
1-2 marks:	One or two ways of increasing the rate are given. No discussion of collision theory is provided.
3-4 marks:	At least two ways of increasing the rate are given. There is some relevant discussion of collision theory. The answer has a logical structure and spelling, grammar and punctuation are mostly correct.
5-6 marks:	Detailed discussion of ways of increasing the rate and relevant collision theory is given. The answer has a logical structure and uses correct spelling, grammar and punctuation.

Here are some points your answer may include:

Collision theory says that the rate of reaction depends on how often and how hard the reacting particles collide with each other. If the particles collide hard enough (with enough energy) they will react.

Increasing the temperature makes particles move faster, so they collide more often and with greater energy. This will increase the rate of reaction.

If the surface area of the catalyst is increased then the particles around it will have more area to work on. This increases the frequency of successful collisions and will increase the rate of reaction.

Increasing the pressure of the hydrogen will mean the particles are more squashed up together. This will increase the frequency of the collisions and increase the rate of reaction.

2 (a) The solid vanadium oxide catalyst gives the reacting particles a surface to stick to *(1 mark)*. This increases the number of successful collisions (and so speeds the reaction up) *(1 mark)*.

(b) Catalysts are expensive to buy *(1 mark)*.
Different reactions require different catalysts so a different catalyst will have to be bought for every reaction that takes place *(1 mark)*.

Page 164

Warm-Up Questions

1) exothermic

2) The temperature will decrease.

3) formed

Exam Questions

1 (a) C—H and O = O *(1 mark)*

(b) 3436 – 2644 = 792 kJ/mol *(1 mark)*

(c) The energy released when the new bonds are formed is greater than the energy needed to break the original bonds, so overall energy is given out *(1 mark)*.

(d)

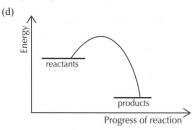

(1 mark for correct axis labels, 1 mark for reactants higher than products)

2 (a)

Time	Temperature of the reaction mixture (°C)		
(s)	1st run	2nd run	Average
0	22.0	22.0	22.0
1	25.6	24.4	25.0
2	28.3	28.1	28.2
3	29.0	28.6	28.8
4	28.8	28.8	28.8
5	28.3	28.7	28.5

(2 marks if all correct, 1 mark for 4 or 5 correct)

(b) (28.8 – 22.0 =) 6.8 °C *(1 mark)*

Don't get caught off guard — the maximum average change is just the highest average temperature minus the lowest average temperature.

(c) Exothermic, because heat is given out to the surroundings/the temperature of the reaction mixture increases *(1 mark)*.

Page 165

Revision Summary for Section 5

13)b)

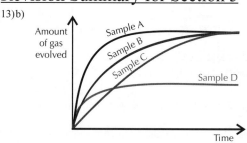

Section 6

Page 171

Warm-Up Questions

1) Neutralisation.

2) A salt and hydrogen gas.

3) Copper nitrate and water.

4) Add the insoluble base to an acid until all the acid is neutralised and the excess base can be seen on the bottom of the flask. Then filter out the excess base and evaporate off the water to leave a pure, dry sample.

5) barium chloride + sodium sulfate → barium sulfate + sodium chloride

Exam Questions

1 (a)

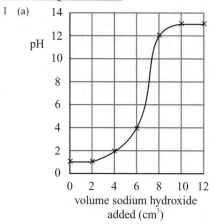

(1 mark for points plotted correctly, 1 mark for best fit curve)

(b) 7 cm³ (also accept answers between 6 and 8, depending on best fit curve at pH 7) *(1 mark)*.

(c) Because the starting pH is pH 1 *(1 mark)*.

This is the kind of question that can somehow trip you up, even if it seems obvious once you know the answer. So it's lucky you've come across it now rather than in the exam, isn't it? The pH before any alkali is added has to be the pH of the acid, and a pH of 1 means a very strong acid. See? Obvious.

(d) sodium sulfate *(1 mark)*

2 (a) It must be insoluble *(1 mark)*.

(b) silver nitrate + hydrochloric acid → silver chloride + nitric acid *(1 mark)*.

(c) First, filter the solution to remove the salt which has precipitated out *(1 mark)*. Then wash the insoluble salt *(1 mark)* and then leave it to dry on filter paper *(1 mark)*.

Page 177

Warm-Up Questions

1) It must be molten or dissolved in water.
2) Oxidation is loss of electrons and reduction is gain of electrons.
3) At the negative electrode.
4) Bromine.
5) The negative electrode.

Exam Questions

1 (a) (i) hydrogen *(1 mark)*

  (ii) $2H^+ + 2e^- \rightarrow H_2$ *(1 mark)*

 (b) (i) chlorine *(1 mark)*

  (ii) $2Cl^- \rightarrow Cl_2 + 2e^- / 2Cl^- - 2e^- \rightarrow Cl_2$ *(1 mark)*

  (iii) E.g. production of bleach / production of plastics *(1 mark)*.

 (c) Sodium is more reactive than hydrogen, so sodium ions stay in solution *(1 mark)*. Hydroxide ions from water are also left behind *(1 mark)*. This means that sodium hydroxide is left in the solution *(1 mark)*.

2 (a) To lower the temperature that electrolysis can take place at *(1 mark)*. This makes it cheaper *(1 mark)*.

 (b) $Al^{3+} + 3e^- \rightarrow Al$ *(1 mark)*

 (c) Oxygen is made at the positive electrode *(1 mark)*. The oxygen will react with the carbon in the electrode to make carbon dioxide *(1 mark)*. This will gradually wear the electrode away *(1 mark)*.

Section 7

Pages 184-185

Warm-Up Questions

1) speed
2) Acceleration — m/s^2, mass — kilograms (kg), weight — newtons (N)
3) $a = (v - u) \div t = (30 - 0) \div 6 = 30 \div 6 = 5 \ m/s^2$
4) gravity

Exam Questions

1 (a) Because its direction is constantly changing. *(1 mark)*

 (b) Acceleration = change in velocity ÷ time taken
   = $(59 - 45) \div 5$
   = $14 \div 5 = 2.8 \ m/s^2$
   (2 marks, allow 1 mark for correct working)

2 (a) (i) 200 m *(1 mark)*
   Read the distance travelled from the graph.

  (ii) speed = distance ÷ time
    $200 \div 15 = 13.3 \ m/s$ (to 1 d.p.)
    (2 marks, allow 1 mark for correct working)

 (b) 13 s (allow answers from 11 s to 15 s) *(1 mark)*
   The bus is stationary between about 33 and 46 seconds.

 (c) The bus is travelling at constant speed (10 m/s) back towards the point it started from. *(1 mark)*

 (d) E.g.

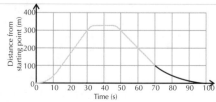

(1 mark)

3 (a) The cyclist is travelling at a constant velocity of 3 m/s *(1 mark)*.

 (b) The cyclist's speed is constantly decreasing. *(1 mark)*

 (c) $(3 - 0) \times (5 - 2) \div 2 = 4.5 \ m$ *(1 mark)*
   Remember, distance travelled is the area under the graph.

4 (a) Weight = mass × gravitational field strength / the weight of an object is proportional to the gravitational field strength *(1 mark)* so the weight of the ball would be less on Mars and so there would be less force pulling the spring down *(1 mark)*.

 (b) $3 \div 1.1 = 2.7$ (to 1 d.p.) So, the spring extends 2.7 times as far on Earth, so g on Earth must be 2.7 times bigger than on Mars. *(1 mark)*
   $10 \div 2.7 = 3.7 \ m/s^2$ (to 1 d.p.)
   (2 marks, allow 1 mark for correct working)

5 (a) Acceleration = change in velocity ÷ time taken
   $= (10 - 0) \div 1 = 10 \ m/s^2$ *(2 marks, allow 1 mark for correct working)*

 (b) Change in velocity = acceleration × time = $10 \times 3 = 30 \ m/s$
   (2 marks, allow 1 mark for correct working)

 (c) Weight = mass × g = $0.12 \times 10 = 1.2 \ N$
   (2 marks, allow 1 mark for correct working)

 (d) None — the stone's acceleration due to gravity is determined by the mass of the Earth, not the stone. *(1 mark)*

6 All of them — everything with mass exerts a gravitational force. *(1 mark)*

Page 191

Warm-Up Questions

1) $(30 + 30) - 10 = 50 \ N$ towards the shore
2) 0 N
3) The acceleration also doubles.
4) Against motion.

Exam Questions

1 (a) (i) $900 \ N - 900 \ N = 0 \ N$ *(1 mark)*

  (ii) Parachutist A is falling at a constant (terminal) velocity. *(1 mark)*

 (b) (i) Terminal velocity is reached when the force of air resistance equals the parachutist's weight. *(1 mark)*
    Weight = mass × g = 70×10 *(1 mark)* = 700 N *(1 mark)*

  (ii) Parachutist A would have a higher terminal velocity *(1 mark)* because he has a greater weight. *(1 mark)* This means the force of air resistance would need to be greater to balance his weight, and air resistance is greater at higher speeds. *(1 mark)*

 (c) Because the parachute increases their air resistance/drag. *(1 mark)*

2 (a) The upwards force must be greater *(1 mark)* because Stefan is accelerating upwards. *(1 mark)*

 (b) $g = 10 \ N/kg$, Stefan's mass is $600 \div 10 = 60 \ kg$ *(1 mark)*
   Force = mass × acceleration = 60×2.5 *(1 mark)* = 150 N *(1 mark)*

3 (a) -500 N. *(1 mark)* If the bat exerts a force of 500 N on the ball, the ball also exerts a force of 500 N on the bat, but in the opposite direction. *(1 mark)*

 (b) The ball's acceleration is greater *(1 mark)* because it has a smaller mass than the bat and receives the same force (F = ma). *(1 mark)*

Pages 197-198

Warm-Up Questions

1) The distance travelled during the driver's reaction time.
2) Braking distance
3) Momentum = mass × velocity (p = m × v)
4) The total momentum before an event is the same as after the event.
5) Because work done is a measure of energy transfer. / Work done and energy transferred are the same.
6) Power is the rate at which work is done.

Exam Questions

1 (a) (i) Accept answers between 12 and 13 m *(1 mark)*

  (ii) 35 m *(1 mark)*

  (iii) 35 m – 12 m = 23 m or 35 m – 13 m = 22 m *(1 mark)*

 (b) Braking distance *(1 mark)*
   Using the graph, thinking distance is about 15 m and braking distance about 38 m.

(c) No, if stopping distance and speed were proportional the relationship between them would be shown by a straight line. *(1 mark)*

2 (a) (i) Momentum = mass × velocity = 100 × 6 = 600 kg m/s to the right
(2 marks, allow 1 mark for correct working)

(ii) Momentum = mass × velocity = 80 × 9 = 720 kg m/s to the left
(2 marks, allow 1 mark for correct working)

(b) (i) Take left as positive, then the momentum of the two players is
720 – 600 = 120 kg m/s. *(1 mark)*
The mass of the two players is 100 + 80 = 180 kg, so the speed is
120 ÷ 180 = 0.67 m/s
(2 marks, allow 1 mark for correct working)

(ii) Left *(1 mark)*
The two players travel in the direction player B was going because player B had more momentum before the collision.

3 (a) 42 000 N × 700 m = 29 400 000 J
(2 marks, allow 1 mark for correct working)

(b) 29 400 000 J ÷ 29 400 N = 1000 m (1 km)
(2 marks, allow 1 mark for correct working)

4 momentum before = momentum after *(1 mark)*

$(1 × 14\ 000) = (1 × –13\ 000) + (235 × v_2)$ *(1 mark)*

$14\ 000 = (235 × v_2) – 13\ 000$

$v_2 = (14\ 000 + 13\ 000) ÷ 235$

= 115 km/s to the right (to 3 s.f.) *(1 mark)*

5 (a) Energy transferred = power × time taken
= 90 000 × 5 = 450 000 J = 450 kJ
(2 marks, allow 1 mark for correct working)

(b) momentum before = 7000 × 5 = 35 000 kg m/s
momentum after = 0
change in momentum = 0 – 35 000 = (–)35 000 kg m/s
(2 marks, allow 1 mark for correct working)
You can ignore the negative sign because you're just asked to calculate the force, not the direction it's acting in.
F = 35 000 kg m/s ÷ 0.5 s = 70 000 N
(2 marks, allow 1 mark for correct working)

Pages 206-207

Warm-Up Questions

1) kinetic energy = ½ × mass × velocity²

2) The energy that an object has because of its vertical position in a gravitational field.

3) There is a big change in momentum over a very short time.

4) E.g. power of the engine, how aerodynamic the car is.

5) elastic potential energy

6) The spring will be permanently stretched.

Exam Questions

1 (a) They help direct the kinetic energy of the crash *(1 mark)* away from passengers to other areas of the car. *(1 mark)*

(b) To increase the time it takes for the person to stop moving *(1 mark)* which reduces the forces acting on the chest. *(1 mark)* They also absorb some of their kinetic energy. *(1 mark)*

(c) Air flows easily over aerodynamic cars so there is less air resistance. *(1 mark)*. Cars reach their top speed when the resistive force equals the driving force *(1 mark)*. With less air resistance to overcome, aerodynamic cars can reach a higher speed before this happens. *(1 mark)*

2 (a) During braking, the vehicle's motor is put into reverse which slows the wheels. *(1 mark)* This motor acts as an electrical generator and converts kinetic energy into electrical energy *(1 mark)* which is stored as chemical energy in the vehicle's battery. *(1 mark)*

(b) The energy transferred by braking is stored rather than wasted, e.g. as heat. *(1 mark)*

3 (a) ½ × 2750 kg × (12 m/s)² = 198 000 J
(2 marks, allow 1 mark for correct working)

(b) The van has more energy as it has a bigger mass. *(1 mark)*

(c) 550 000 J ÷ 25 m = 22 000 N
(2 marks, allow 1 mark for correct working)

4 (a) G.P.E. = m × g × h = 2.5 kg × 10 N/kg × 1.3 m = 32.5 J
(2 marks, allow 1 mark for correct working)

(b) K.E. = ½ × m × v², so v² = K.E. ÷ (½ × m)
v² = 32.5 J ÷ (½ × 2.5 kg)
v = 5.1 m/s (to 1 d.p.)
(3 marks, allow 1 mark for correctly rearranging the equation and 1 mark for correct substitution of values into the equation)

(c) F = k × e
k = F ÷ e = 4 N ÷ 0.035 m = 114 N/m (to 3 s.f.)
(3 marks, allow 1 mark for correctly rearranging the equation and 1 mark for correct substitution of values into the equation)

5 (a) Doubling the mass of the car will double the kinetic energy of the car at a particular speed *(1 mark)*. As K.E. = F × d, the braking distance will also double when the mass of the car is doubled *(1 mark)*.

(b) ABS brakes are safer to use as they help drivers keep control of the car's steering when braking hard *(1 mark)*. They automatically pump on and off to stop the wheels locking and prevent skidding *(1 mark)*.

(c) Heat *(1 mark)*
Sound *(1 mark)*

Page 208

Revision Summary for Section 7

5) a = (v – u) ÷ t; 35 m/s²

11) F = ma, so a = F ÷ m; 7.5 m/s²

12) 1.33 m/s²

13) 120 N

16) W = F × d; 6420 J

17) KE = ½ × m × v² = 20 631 J

18) The car would stop in 5.1 m, so he will hit the sheep.

19) a) 1200 J

b) 600 J

20) 15 600 J

Section 8

Pages 213-214

Warm-Up Questions

1) Insulator

2) Positive and negative

3) Repel

4) E.g. bike / car

Exam Questions

1 (a) A *(1 mark)* The rod is negatively charged so would repel the negative charges in the balloon, making them move away from the rod *(1 mark)*.

(b) The negative charges in the rod attract the positive charges in the balloon *(1 mark)*.

2 (a) Electrons are scraped from the cloth onto the surface *(1 mark)*.

(b) -23 *(1 mark)*

(c) E.g. copper *(1 mark)*.

3 (a) (i) Jyoti might get an electric shock from the chair leg *(1 mark)*

(ii) Jyoti's shoes insulate her from Earth so she will become statically charged when she walks on the carpet *(1 mark)*. This charge will flow to earth when she touches the conducting material *(1 mark)*.

(b) (i) Touching the metal chair leg will connect Jyoti to Earth and allow any charge that has built up on her to disperse. *(1 mark)*

(ii) No, because wood is an insulating material so would not allow the charge to flow to Earth *(1 mark)*.

4 Rubbing the bristles against the cloth will cause a build up of static electricity, *(1 mark)* which will attract the gold leaf *(1 mark)*.

5 (a) Restarting people's hearts *(1 mark)*.

(b) Insulating, so that the only person to be given a shock is the patient *(1 mark)*.

(c) E.g. paint sprayer / dust precipitator. *(1 mark)*

6 (a) As fuel flows out of a filler pipe static could build up *(1 mark)*. This built up charge could cause a spark that might cause an explosion. *(1 mark)*

(b) An earthing cable provides an easy route for the static charges to travel into the ground *(1 mark)*. This means no charge can build up and cause a spark *(1 mark)*.

Page 221

Warm-Up Questions

1) Ohms, Ω

2) Potential difference = work done ÷ charge (V = W ÷ Q)

3)

4) E.g. automatic night lights / outdoor lighting / burglar detectors.

5) It decreases.

Exam Questions

1 (a) (i) Resistance = potential difference ÷ current = 1.5 ÷ 0.3 = 5 Ω
(2 marks, allow 1 mark for correct working)

(ii) Charge = current × time = 0.3 × 35 = 10.5 C
(2 marks, allow 1 mark for correct working)

(b) The amount of current flowing will decrease *(1 mark)*.

(c) (i)

(1 mark)

(ii) As electrical charges flow through the filament some of the electrical energy is transferred into heat energy *(1 mark)*. This causes the ions in the filament to vibrate more *(1 mark)* which makes it harder for the charges to move through the filament, so the resistance increases *(1 mark)*.

2 (a) Diodes only allow current to flow in one direction. *(1 mark)*

(b) Resistance = potential difference ÷ current = 6 ÷ 3 = 2 Ω
(2 marks, allow 1 mark for correct working)
Read the values from the graph, then use the formula R = V ÷ I.

(c) E.g. They use a smaller current *(1 mark)*.

Page 227

Warm-Up Questions

1) The total potential difference of the cells and the total resistance of the circuit.

2) Parallel

3) 50 Hz

4) Direct current (d.c.)

5) 1 ÷ 100 = 0.01 s

Exam Questions

1 (a) Total resistance = $R_1 + R_2 + R_3$ = 2 + 3 + 5 = 10 Ω
(2 marks, allow 1 mark for correct working)

(b) The current will be 0.4 A *(1 mark)* because in a series circuit the same current flows through all parts of the circuit *(1 mark)*.

(c) $V_3 = V - V_1 - V_2$ = 4 − 0.8 − 1.2 = 2 V
(2 marks, allow 1 mark for correct working)

2 (a) The trace shows an AC source so cannot be from a battery / must be from mains electricity *(1 mark)*.

(b) 20 ms *(1 mark)*
The wave takes four divisions to repeat. 4 × 5 ms = 20 ms.

(c) 20 ms = 0.02 s. Frequency = 1 ÷ time = 1 ÷ 0.02 = 50 Hz
(2 marks, allow 1 mark for correct working)

(d) The amplitude (vertical height) of the wave will be decreased so the peaks and troughs will be smaller *(1 mark)*.

3 (a) 15 V *(1 mark)*
Potential difference is the same across each branch in a parallel circuit.

(b) Current = potential difference ÷ resistance = 15 ÷ 3 = 5 A
(2 marks, allow 1 mark for correct working)

(c) 5 + 3.75 = 8.75 A
(2 marks, allow 1 mark for correct working)

Page 234

Warm-Up Questions

1) The neutral wire (also accept the earth wire).

2) Plastic is a good insulator.

3) Earth wire

4) Electrical energy to heat energy.

5) E (energy) = Q (charge) × V (voltage).

Exam Questions

1 (a) (i) brown *(1 mark)*

(ii) blue *(1 mark)*

(iii) green and yellow stripes *(1 mark)*

(b) the live wire *(1 mark)*

(c) a fuse *(1 mark)*

2 How to grade your answer:

0 marks:	There is no relevant information.
1-2 marks:	There is a brief description of how the earth wire and fuse protect the appliance and prevent electric shocks.
3-4 marks:	There is some description of how the earth wire and fuse protect the appliance and prevent electric shocks. The answer has a logical structure and spelling, punctuation and grammar are mostly correct.
5-6 marks:	There is a clear and detailed description of how the earth wire and fuse protect the appliance and prevent electric shocks. The answer has a logical structure and uses correct spelling, grammar and punctuation.

Here are some points your answer may include:

- The earth wire and fuse are used to protect the circuit from being damaged by current surges.
- The metal case of the appliance is earthed using the earth wire.
- If the live wire touches the metal case, a huge current will flow through the live wire, through the case and then out through the earth wire.
- This surge in current melts the fuse, which breaks the circuit and cuts off the electricity supply. This isolates the whole appliance and protects the circuits and wiring in the appliance from damage.
- Isolating the appliance also makes it impossible to get an electric shock from the case.
- Shutting off the live supply also prevents fires caused by the heating effect of a large current.

3 (a) Power = current × voltage = 0.5 × 3 = 1.5 W
(2 marks, allow 1 mark for correct working)

(b) Charge = current × time = 0.5 × (30 × 60) = 900 C
(2 marks, allow 1 mark for correct working)

(c) Energy = charge × voltage = 900 × 3 = 2700 J
(2 marks, allow 1 mark for correct working)

Pages 242-243

Warm-Up Questions

1) The creation of a voltage across a conductor by moving a magnet in or near a coil of wire.

2) More turns on the secondary coil.

3) $V_p/V_s = N_p/N_s$ or $V_s/V_p = N_s/N_p$

4) First finger — field, Second finger — current, Thumb — motion

5) Any three from, e.g. DVD players / electric cars / electric trains / hard disk drives / fans / washing machines / fridges / vacuum cleaners.

Exam Questions

1 (a) As the wheel moves round it turns the cog wheel which is attached to a magnet *(1 mark)*. The magnet rotates near a coil of wire *(1 mark)*. This creates a voltage/current in the wire which is used to power the light *(1 mark)*.

   (b) (i) No voltage is induced. / The output drops to zero *(1 mark)*.

       (ii) The lights would go out when the rider stops moving, e.g. at a junction *(1 mark)*

2 (a) It travels from the north pole to the south pole (from left to right) *(1 mark)*.

   (b)

(1 mark)

   (c) The direction of the force would be reversed too *(1 mark.)*

   (d) (i) The force would be less *(1 mark)*.

       (ii) The would be no force on the wire *(1 mark)*.

3 (a) anticlockwise *(1 mark)*
     Use Fleming's left-hand rule on one of the arms of the coil.

   (b) The split-ring commutator keeps the motor turning in the same direction by swapping the contacts every half turn *(1 mark)*.

   (c) E.g. increase the current flowing through the coil / use stronger magnets / add an iron core to the coil / increase the number of turns on the coil *(1 mark)*.

   (d) E.g. by reversing the direction of the current / the magnetic field *(1 mark)*.

4 (a) The transformer will have more turns on its primary coil because it is a step-down transformer *(1 mark)*.
     Step-down transformers have fewer turns on the secondary coil to reduce the voltage.

   (b) The primary coil produces a magnetic field which stays within the iron core *(1 mark)*. Because there's an alternating (AC) in the primary coil, the magnetic field in the iron core is constantly changing direction *(1 mark)*. This changing magnetic field induces an alternating voltage in the secondary coil by electromagnetic induction *(1 mark)*.

5 (a) A voltage will not be induced in the secondary coil when using the DC supply from the battery *(1 mark)* because the magnetic field generated in the iron core is not changing *(1 mark)*.

   (b) (i) Power = current × voltage = 2.5 × 12 = 30 W
         (2 marks, allow 1 mark for correct working)

       (ii) Assuming that the transformer is 100% efficient, the output power is 30 W *(1 mark)*.
         Current = power ÷ voltage = 30 ÷ 4 = 7.5 A
         (2 marks, allow 1 mark for correct working)

       (iii) $V_s \div V_p = N_s \div N_p$
          $4 \div 12 = N_s \div 15$
          $60 \div 12 = 5$ turns
          (2 marks, allow 1 mark for correct working)

Page 244

Revision Summary for Section 8

4) 4 A

10) 1.2 V

11) 4.8 W

16) 12.5 Hz

21) Hair straighteners: E = 13 500 J
    Hair dryer: E = 12 600 J
    The hair straighteners use more energy.

22) 3180 J

25) $V_p \div V_s = N_p \div N_s$ so
    $V_p = (500 \div 20) \times 9 = 25 \times 9 = 225$ V

Section 9

Page 251

Warm-Up Questions

1) Electrons

2) E.g. fallout from nuclear weapons tests / nuclear accidents / dumped nuclear waste / medical x-rays.

3) E.g. location and job.

4) Cosmic rays

5) Alpha particles

Exam Questions

1 (a) (i) -1 *(1 mark)*

      (ii) +1 *(1 mark)*

      (iii) 0 *(1 mark)*

   (b) Protons *(1 mark)* and neutrons *(1 mark)*

   (c) It increases by one. *(1 mark)*

   (d) It decreases by four. *(1 mark)*

   (e) (i) The number of protons and neutrons in the atom. *(1 mark)*

      (ii) Atom A and atom B *(1 mark)* because isotopes of the same element have the same atomic number but different mass numbers. *(1 mark)*

   (f) (i) They have opposite charge. *(1 mark)*

      (ii) Alpha particles have a much greater mass. *(1 mark)*

2 (a) Most of the alpha particles went straight through the foil. *(1 mark)* But a small number of alpha particles were deflected straight back at them. *(1 mark)*

   (b) E.g. Most of the atom is empty space. *(1 mark)* The nucleus of an atom is tiny *(1 mark)* and contains most of the mass *(1 mark)* and is positively charged. *(1 mark)*

Pages 258-259

Warm-Up Questions

1) Because it is ionising and can damage cells.

2) Being exposed to radiation without coming into contact with the source.

3) Lead has radiation absorbing properties — it absorbs all three types of radiation (alpha, beta and gamma) .

4) A weak alpha source is placed in the detector close to two electrodes. The source causes ionisation of the air particles which allows a current to flow. If there is a fire, then smoke particles are hit by the alpha particles instead. This causes less ionisation of the air particles — so the current is reduced causing the alarm to sound.

Exam Questions

1 Smoke detectors *(1 mark)*

2 (a) Alpha radiation is stopped by the body's tissues and so wouldn't be detected externally. *(1 mark)* It is also strongly ionising which makes it dangerous inside the body. *(1 mark)*

(b) So that the device lasts a long time and therefore doesn't need to be replaced as often. *(1 mark)*

(c) So that the dose to the rest of the body is minimised / to reduce damage to healthy cells. *(1 mark)*

3 (a) The radiation can break up molecules in the body's cells into reactive ions which can damage the cells without killing them, causing cancer *(1 mark)*.

(b) It can kill cells, which causes radiation sickness *(1 mark)*.

4 (a) The average time it takes for the number of nuclei in a radioactive isotope sample to halve / the time taken for the count rate or activity to halve. *(1 mark)*

(b) one quarter / 25% *(1 mark)*

(c) E.g. any two from: keep exposure time short / don't allow skin contact with sample / hold container at arm's length / wear protective lead clothing / put it in a lead container / avoid looking directly at the sample. *(1 mark for each)*

5 (a) 1 = alpha
2 = beta
3 = gamma *(1 mark)*

(b) Alpha radiation is the most dangerous type of radiation inside the body *(1 mark)* because it does all its damage in a localised area *(1 mark)*.

6 (a) the thyroid gland *(1 mark)*

(b) gamma radiation *(1 mark)*

7 (a) No. If the source's half-life is very short, it will decay very rapidly and so need to be changed very frequently *(1 mark)*.

(b) It doesn't need high temperatures *(1 mark)*, so delicate/plastic equipment can be sterilised without being damaged *(1 mark)*.

8 (a) Alpha radiation would be completely stopped by paper *(1 mark)*.

(b) Gamma radiation would not be blocked by even thick paper and so couldn't provide any information about thickness *(1 mark)*.

Pages 265-266

Warm-Up Questions

1) E.g. uranium and plutonium.

2) It produces a lot of radioactive waste that must be carefully disposed of.

3) a) It would have made it possible to generate a lot of electricity easily and cheaply.

b) The results have not been repeated reliably enough.

4) Clouds of dust and gas.

5) No — our Sun is a small star. Only big stars become black holes.

6) A red giant.

Exam Questions

1 (a) (i) 'Low level' means that the waste is only slightly radioactive *(1 mark)*.

(ii) By burying it in a secure landfill site *(1 mark)*.

(b) If the site is not geologically stable there could be earthquakes *(1 mark)*. Earthquakes could break the canisters that the nuclear waste is stored in which could allow nuclear waste to leak out *(1 mark)*.

2 (a) Deuterium *(1 mark)* and hydrogen *(1 mark)*
Fission uses heavy elements, whereas nuclear fusion uses light elements.

(b) Fusion power would allow a lot of electricity to be generated from a plentiful fuel *(1 mark)* without the large amounts of waste currently produced by fission. *(1 mark)*

(c) Fusion only works at such high temperatures and pressures that it uses more energy than it can produce. *(1 mark)*

3 (a) Stars form from clouds of dust and gas which spiral in due to gravitational attraction. *(1 mark)* Gravitational energy is converted into heat energy so the temperature rises. *(1 mark)* When the temperature gets hot enough, nuclear fusion happens and huge amounts of heat and light are emitted. *(1 mark)*

(b) The heat caused by nuclear fusion provides an outward force to balance the force of gravity pulling everything inwards. *(1 mark)*

(c) (i) They become unstable and eject their outer layer of dust and gases as a planetary nebula *(1 mark)* which leaves a hot, dense solid core known as a white dwarf. *(1 mark)* White dwarfs then cool to become black dwarfs. *(1 mark)*

(ii) They start to glow brightly again and undergo more fusion, and expand and contract several times. *(1 mark)* Heavier elements are formed and the star eventually explodes in a supernova. *(1 mark)* The supernova leaves behind a neutron star. *(1 mark)* If the star is big enough this will become a black hole. *(1 mark)*

4 (a) (i) heat *(1 mark)*

(ii) steam *(1 mark)*

(b) (i) mass number of uranium + 1 neutron = mass number of krypton + mass number of barium + 3 neutrons
235 + 1 = mass number of krypton + 144 + 3
mass number of krypton = 235 + 1 − 144 − 3 = 89
(1 mark for correct answer, 1 mark for correct working)

(ii) U-235 is bombarded with slow-moving neutrons *(1 mark)*. A U-235 nucleus absorbs a neutron *(1 mark)*, splits into two smaller nuclei (Kr and Ba) and releases 3 neutrons *(1 mark)*. These neutrons can go on to collide with other uranium atoms, creating a chain reaction *(1 mark)*.

Page 267

Revision Summary for Section 9

10)(a) $^{131}_{53}\text{I} \rightarrow {}^{131}_{54}\text{Xe} + {}^{0}_{-1}\text{e}$

(b) $^{148}_{64}\text{Gd} \rightarrow {}^{144}_{62}\text{Sm} + {}^{4}_{2}\text{He}$

12) 1 hour 20 minutes

Index

Index

Index